FAST CYCLE PRODUCTION

The Manufacturing Philosophy That Always Works

By

Tom Clason

ISBN: 1-4107-4865-0 (Paperback)

This book is printed on acid free paper.

1st Books - rev. 12/02/03

Preface

For decades I have been convinced that **any** company could achieve improved productivity if only they would base each and every decision that they make on a proper set of core values. My biggest challenge throughout my consulting career has been to educate my associates and clients in what those values were and why they were so important to the achievement of productivity improvement.

As we all know, there are plenty of consultants out there that work to help their clients improve their ability to make a profit. My area of expertise is specifically designed to help clients produce more products with fewer resources. This book will not help a business to improve their sales technique, but it certainly will help them improve their sales through competitive ability. If they have a good product, this book will help them to improve their ability to get their product to their customers on time, with better quality and in the amounts that the customers demand. The process should be every bit as competitive as the product is! This book will help the readers to improve their processes.

The world of production is a tough one. It's tough to even stay alive in this business, let alone to prosper. With such a tough industry, it has always been amazing to me that a good percentage of production managers at every level know very little about the science of production. They might know the technical details of their production process, but in many cases, they are dangerously naïve when it comes to their processes. Many of them have seriously misguided production values and it is apparent whenever they get into a crunch. When the situation is desperate; when there is more demand than capability, these same managers make their most dangerous mistakes. They do not make mistakes because they are stupid; they make mistakes because they are not schooled in the science of manufacturing.

It is my hope that this book will help production managers out there to establish a solid understanding of the **art** and science of manufacturing. Yes, I did say the "art and science" of manufacturing. The ability to apply good, solid values to the variety of issues and problems that arise each and every day, and to always come up with solutions that are consistent with those values is an art form in itself. It is not necessary to remain unenlightened any longer. The decision process may require deeper thought, but the end result of doing the right things will pay huge dividends both short term and long.

Believing in sound production values and consistently making decisions that are consistent with those values is what this book is all about. Throughout this book, I refer to "Values Based This, and Values Based That. I speak of Values Based Manufacturing, Values Based Management, Values Based Run Rules and Values Based Production Tools. Values based decision making is the foundation on which this book is built. I believe strongly that if all solutions to production issues and problems are foundationally based on good, solid production values, those decisions will have a positive effect upon the ability of the company to prosper both short and long term.

I have dedicated a good portion of my adult life in the quest to design a productivity improvement system that will never fail. I believe that Fast Cycle Production is that system. The Fast Cycle Production system can and will help any producer dramatically improve their production capability without increasing resources, which is what true productivity is all about. If growth becomes necessary, and it most certainly will be necessary if there is a strong product, the company will have an excellent chance of growing intelligently if the process changes that it makes are based on sound production values. Adopting Fast Cycle Production values and tools will make the transition from small to large relatively painless. The Fast Cycle Production Values Based system will not only ensure intelligent growth, but will sustain the improvements for the long term. Long-term productivity with growth is only possible when the growth is realized with the minimum amounts of increased resources and capital expenditures possible.

Those Who Made It All Possible

I could not have realized the success that I've achieved without the help of a special group of people that believed in me. Several past IBMers that I am especially grateful to are: Mike Hills and Robert Owen, both of whom were successful IBM managers when I first met them and eventually left IBM for new opportunities elsewhere.

Mike was the IBM Project Manager for the CFM Core Team at IBM Austin, a panels manufacturing plant when we first met. Mike didn't just accept Takt as a management tool, but embraced it. Neither of us knew it at the time, but that project would be the beginning of TAKT Management. The results achieved with FCP were phenomenal and actually saved the factory from folding. Mike and I worked together very closely during the project in Austin. Mike has been one of my strongest advocates through the years and has remained a friend. Because of him, I was able to prove that Takt was not just an implementation tool. Mike gave me the authority to implement my own version of setup reduction at Austin, which was a huge success. By giving me the authority to modify the CFM implementation curriculum, Mike demonstrated a great deal of faith in me. He believed in me and felt that he was doing the right thing for his company. Mike believed in me so much, that after he moved to Endicott, he recommended me for the Endicott project to Bob Owen. I will be forever grateful and loyal to Mike for the trust that he has shown in me throughout the years.

Bob Owen, then the Manager of Manufacturing Operations for Endicott Panels Plant, showed a rare form of courage when he chose our firm over an internal consulting group in competition for a major consulting contract. Bob was put under tremendous pressure by a Senior Executive for choosing us to help his company return to profitability and productivity. At the time, the Panels Plant in Endicott, New York was suffering both operational and financial trouble. They were operating at a loss and were losing customers. Bob was told that there might be repercussions if he chose our firm instead of the internal consulting group. Bob believed that we would provide him with the highest probability of success and he held his ground. He was doing what he believed was best for his company, and he was not about to be intimidated by potential threats to his career. I have found that brand of courage to be rare in the business world. To Bob I owe my undying admiration, gratitude, loyalty and friendship.

Gerry Clarke believed in me also. Clarke was my partner in the previously mentioned contract with IBM Endicott. Clarke and I put together the plan for Endicott that completely reversed their productivity and cultural dilemmas within a year. We accomplished a lot and we worked hard together. There were those that referred to us as "Tom and Gerry", partly because our names fit the bill, and partly because we sometimes fought like cats and dogs. But through all the turmoil, I knew that I could depend upon Clarke. He had wisdom and a patience that I have never seen in a person before we met. He was always patient with me and was the perfect partner and friend. I miss working with Clarke. After our last project together, he went on to become the president of Human Synergistics, Inc, which is a very successful leadership consultant company that specializes in organizational culture transformation. Clarke helped me to improve as a person, and a consultant. For that and much more, I will always consider him a valued friend. I look forward to someday having a drink or ten with Clarke and talking about projects past.

Most important of all, I want to thank my wife, Marlene for being my strongest supporter for the last forty-four years and counting. She has believed in me when I doubted myself. Without her, nothing I've accomplished would have been possible. Whenever I was out of town, which was often, she faithfully maintained the household. When I returned, she was there to greet me with love and caring. She is the CFO for our family and our business finances, which she capably manages with the no-nonsense mentality of an IRS auditor. I have already committed my life and love to her for all time, so I will simply dedicate this book to her. Thank you my love, this book is dedicated to you for all that you have done and will continue to do.

Forward

The book, **"Fast Cycle Production: The Manufacturing Philosophy That <u>Always</u> Works!"** is a comprehensive summary of the most powerful productivity tools available to the business world. The author, Tom A. Clason, brilliantly outlines his winning approach to Manufacturing Excellence, which he has been using for several decades to help companies gain market share and defeat their competition. As a successful consultant to a variety of leading companies, Tom Clason has demonstrated that his approach to manufacturing competitiveness works. *Fast Cycle Production* is as good if not better than any other cycle time reduction technique being offered today, and more importantly, his tools allow these competitive gains to be continuously improved upon. In today's challenging Manufacturing environment, every company is looking for an advantage over its competition. *Fast Cycle Production* allows for extraordinary gains to factory throughputs, as well as helping to achieve dramatic cycle time and WIP reductions. These gains are realized in the order of just a few months, as compared with other approaches that may take years to achieve comparable results. And while maintaining the gains is a challenge for any continuous improvement process; Fast Cycle Production provides the tools for sustaining an on-going operational improvement effort.

I first met our author during a Fast Cycle Production implementation at a Fortune 50 company and a leading computer manufacturer for which I worked. My 15 plus years of experience prior to leading this FCP implementation (under the direction of Tom Clason, of course), hadn't quite prepared me for the operational transformation I would experience. Previously, I held numerous positions for leading Manufacturing companies, and served in roles such as Director of Manufacturing, Operations Manager, Plant Manager and Corporate Vice President. During these assignments, I was responsible for many operational improvement initiatives, and lead several *Lean Manufacturing* and *kaizen* activities. Despite my manufacturing background, I had never experienced as comprehensive an offering as the Fast Cycle Production approach.

Tom Clason's FCP techniques embody the Art and Science of Manufacturing as no other approach that I've seen.

As the FCP Implementation Team leader, I was first struck by Tom Clason's passion for the Fast Cycle Production process. His beliefs are deeply rooted in Fast Cycle Production's values and principles, and his vigorous passion is reflected in this book. Tom Clason captures decades of wisdom and experience that should be invaluable to many in the Manufacturing and Service industry. The recent migration of manufacturing jobs outside the US borders should be a concern for all of us. My sincere hope is that organizations will embrace this approach and apply it in their own organizations. These methods are key to achieving Manufacturing Excellence and will be key in maintaining the competitive industrial advantage that we have worked so hard to achieve in this country.

Competing in today's marketplace requires superior manufacturing strategies and a constant vigilance towards increasing productivity by improving production processes. Fast Cycle Production's time-tested approach to process improvement has helped a countless number of companies, including several Corporate 100 companies. By using the approaches outlined in this book, companies have pulled themselves out of pending failure, and turned their organization into industry leaders. This easy to follow approach to achieving World Class manufacturing status becomes available to everyone through this powerful book. "Fast Cycle Production" provides the answers to your organization's productivity needs, whether you're interested in maintaining your competitive edge, or need to completely reinvent your manufacturing processes, these techniques can help you achieve dramatic results. When used as Tom Clason prescribes for his readers, these principles <u>always</u> produce winning results!

FCP is a manufacturing philosophy that has as its foundation in the principles first employed by the champions of the Toyota Production System. While similar to some of the approaches found in the *Lean Manufacturing* technique, Fast Cycle Production has a very unique approach and provides very powerful productivity tools that help companies realize phenomenal gains. Fast Cycle production draws upon the best of the "Manufacturing Masters". Productivity principles from heroes like; Shingo, Goldratt, Deming and others are at the foundation of these timeless techniques. FCP blends these approaches masterfully into its own unique continuous improvement process.

Fast Cycle Production is the "manufacturing philosophy that <u>always</u> works". Value-Based Manufacturing (VBM) reinforces your manufacturing sciences through an acceptance and understanding of the values and principles utilized by world-class producers. VBM is the foundation upon which the Fast Cycle Production processes is built, and will provide your organization a platform to launch the most powerful continuous improvement tool ever. If you are presently utilizing or are planning to implement Continuous Flow Manufacturing, The Toyota Production System or Lean Manufacturing techniques in your company to improve your business, your effectiveness and longevity will become severely compromised without a VBM approach to continuous process improvement. The effective use of Manufacturing Run Rules will help your business incorporate Fast Cycle values and principles into all of its critical business processes.

Agility and flexibility within the organization are also achieved through these processes. Many organizations recognize that in order to compete and stay competitive, it must be flexible enough to respond quickly and effectively to a changing market or industry. Agility is becoming a critical competitive advantage for any organization, and in particular a manufacturing company. The JIT techniques, which Tom Clason, outlines provide for tremendous manufacturing flexibility. His TAKT Management process is essential for providing effective use of resources and information…both of which are crucial in becoming agile. An organization whose Culture is prepared to embrace these changes in customer needs and market demands will realize tremendous gains from these tools.

Kaizen, or continuous improvement, is sustained through Fast Cycle Production's unique approach to a company's TAKT (or the BEAT) of the process. Through the use of its TAKT Management process, Employee Empowerment and Continuous Process Improvement are realized throughout the production floor. The extremely powerful TAKT Management system provides an approach that helps a company realize meaningful operational improvements and ensures that it will sustain these gains. The TAKT Management process is recognized by its users as the most powerful productivity tool ever!

Other important improvement techniques utilized by the Fast Cycle Production process include: Kanbans (and FCP's unique Dynaban approach), Gating-Process Folding, Group Technology, Single-Unit Flow, Circles of Motion, Process and Product Mapping, Value-Add Analysis, Opportunity Analysis and Simulation Modeling. In a matter of just a few weeks, these productivity tools will help an organization achieve dramatic results. Improvements such as increased factory throughputs, Cycle Time reduction, Inventory and WIP reduction, Floor Space optimization, increases in Outputs from gating operations, and improvements in production Efficiencies are all benefits in a Fast Cycle Production factory. If your business can benefit from these types of dramatic improvements, then Fast Cycle Production's time-tested technique is the continuous improvement process for you. Any business owner, CEO, or Operations Manager will find their valuable time well spent by reading this powerful book and applying the Fast Cycle Production principles in their own organizations.

January 31, 2003   Michael V. Testani, Sr.
Process and Manufacturing
Engineering Manager
Complex Assembly Operations
Microelectronics Division
IBM Corporation

Part 1 – The Birth Of The Belief

Tom Clason

Chapter 1 - In The Beginning ...

In this section you will learn how Fast Cycle Production began, who influenced my philosophy and how I formed my methodology for the implementation of my system. For the most part I try to keep it in chronological order, but I do jump around a bit to make or emphasize certain points. It is my intent in this section to show why I believe in Fast Cycle Production so very much and how I formed my Values Based Management methodology.

Before I start, I want you to know that although I have modified the use of many of the tools that were initially developed and used by others, the basic core values are pretty much the invention of a very distinguished group of pioneers in the fields of Just In Time (JIT) and Continuous Flow Manufacturing (CFM), along with innovators like Deming and Goldratt whose brilliant approaches to process excellence paved a path for many of us to follow. It would be very gratifying to be able to say that I am the inventor of the values and tools that I now believe in, but I am not. It would be impossible to improve on many of the values already established, simply because they are as sound today as any out there and have proven their worth over time.

I have modified the application of most of the tools that are contained within; some I have modified dramatically, others less so. You might say that I have modernized them to make them more useful in today's production environment. There are a few tools that I will introduce to you that I can say were of my own design – Dynabans and Circles of Motion are examples. I can say with great assurance that I am a true and vocal believer in what I profess. I work diligently to convince others to do the same. My strong beliefs are the reason that I set out to write this book. Those that know me know of my passion for the achievement of process excellence. No matter how hard the challenge, they know that I will stand behind what I believe. I never wilt under pressure.

I did not originally set out to create a system that could be taught to others. When I first started, I was just trying to do my very best to be a successful manufacturing manager. Fighting daily battles to improve my own process was my focus. There were no lofty ideals when I first started, and some of my present beliefs were the result of trial and error, not intellectual endeavor. I used my best judgment at the time to arrive at solutions to the problems that I experienced. Unfortunately, I had no basic values from which to base those decisions. Originally, I was "flying by the seat of my pants" and hoping for the best.

In my search for knowledge, I began to study the published works of the creators of JIT (Just In Time) Manufacturing. The more I learned, the more I wanted to try to bring JIT into my own company. I was able to learn everything about JIT except how to implement it. That is where CFM (Continuous Flow Manufacturing) gave me a great deal of help. But the CFM implementation methodology had some weaknesses that I would one day be determined to improve upon. It is exactly that which led me to search for a better implementation system. It was not my intent initially to invent a completely new and different system. That came much later.

I learned that there are some very appropriate questions that should be considered when implementing a new system into a factory. What comes first, second, third and so on? What should the training program look like? Along with the implementation team, should you train a few key employees, or everyone? What tools, like Kanbans, TAKT and others should you use and under what conditions? What is contained in a process mapping effort, and should you always include process mapping? What about simulation modeling? When do you use it and when do you not use it? Where is single unit flow used appropriately? What is the proper batch size for a machine? How do you go about doing setup reduction in a plant?

These are some of the questions that are appropriately asked whenever a process simplification and improvement effort is underway. There are many more. My work over the years has been to answer all of these questions, or at least give guidelines that would allow the implementers to answer the questions when they surfaced. Without guidelines and process values to guide you, all you can do is try this, try that, and hope that you are making the right moves.

But as any reasonable person knows, trial and error does not always work. There are times that a serious mistake can result in a manager or even a company finding itself in very serious if not terminal trouble. It is so much better to find and stick with a system that has a solid reputation, than to act like a loose cannon. It is my hope that many others will avoid the pitfalls of "dice rolling" decisions. I do not care how good your product is, gambling with your process is not a good way to beat the competition. In some cases, getting your product to market quickly and efficiently can make the difference between success and failure. If you do not get your product to market quickly, and with excellent quality, you just might not stick around very long. Even if you can survive, you will not make the profit that you could and should have made if you were just better at what you do.

The Corporate Standard

The standard method of controlling costs has traditionally been the easiest: Cut the cost of labor and cut the cost of spending. Those are the easiest variables to work with that will improve your bottom line. Some CEO's have made huge bonuses by doing nothing more than cutting back labor. In many cases the cuts have been indiscriminately administered with little concern for process excellence. The cuts were most likely surgical, and in many cases, were probably not the best long-term approach. They were merely the easiest.

I do not mean to insinuate or state that those CEO's were highly paid empty suits. They did not get where they are by being stupid, or inept. But I have learned over time that some of the most successful of them are very capable of making very questionable decisions. In many cases they make their move because of outside pressure - the stock market, for example. Other times they might make their decisions based on input from others and their own beliefs and observations. Whatever their reasons for acting, their solution to their problem usually follows the path of least resistance, working with variables that require the least amount of planned thought, such as downsizing the work force through layoffs and/or attrition.

If they do decide that downsizing is necessary, you can bet that they will not be the ones who will carry out the details. Their job is to make the order, not execute the implementation of it. Once the order is made, it is their subordinate managers who will do the dirty work. That is as it should be I guess. I do not have a problem with them making their decisions - that is what they are paid to do. What I have

5

a problem with is that there is not often enough consideration given to the effect that the executive decision will have on the **process**.

When a company has a great product and the demand for it is strong, staffing up to meet the demand is an area where senior and middle managers also miss the boat. Because they have such a strong product and the orders are coming in like crazy, the chaotic methods they use to increase (?) their capabilities are often the reasons that consultants like me are called in to help straighten out the mess. They just throw money and people at the problem, without worrying about what the true consequences will be.

New machines are purchased and put wherever there is space. There is no pause or planning to make sure the machines are put where they will function the best. I have worked with clients who put their new machines so far away from where they should be placed that any potential increased capability is lost because there is so much transport of product required.

Buy the machines and put them wherever you can. Don't worry about hurting the process. Hire as many people as you think you need. Don't worry about the fact that doubling your work force **does not** result in a doubling of your capability. Lay people off when you don't need them, even though you may have to rehire soon afterwards. Better still; build up an excessive temporary work force so you won't have to worry about the cost of employee benefits. Put those temporary workers in areas requiring a great deal of skill. Don't put them in areas requiring very little training. When they leave, and they probably will, you will have to go to the great expense of training new people and temporarily losing output capability because the new people can't come up to speed quickly enough. Don't worry about the future, and don't worry about establishing and maintaining a skilled, dedicated work force. That requires too much planning and analysis. Only worry about the short-term cost. I have seen all of these examples and they are all vehicles for failure.

Whenever a company's order volume deteriorates and/or costs rise, the scenario is set for disappointing earnings. Sometimes a company has more orders than they are able to produce resulting in missed windows of opportunity. As soon as a company is unable to "beat the street," their stock will likely suffer in the market. Once this happens, my guess is that the probability is high that there will be layoffs. Why will there be a layoff, and not some other response?

Because laying people off works when you want to impress Wall Street. If you fail to do something to cut costs quickly, you may not beat the street. If you don't beat the street, your stockholders will put pressure on you. You can almost call this an unfortunate reality of life in the world of production.

I know of no tougher business to be in than manufacturing. You are up against insurmountable odds. Everyone out there is trying to take market share from you and you must hold your costs down and still come up with a quality product. At times you find that you have created something special and customers are lining up to buy what you are selling. Suddenly, you find that you could sell much more if only you could produce it. You throw money at your process in order to meet product demands and what do you get? You get to spend money.

In large companies with multiple plants, it is the plants that are not performing well that will come under pressure. There are usually a variety of **administrative** solutions that are thrown in to relieve the pressure. The next thing that is tried is the systematic use of extreme pressure on certain people in the plant. Meetings are held daily, twice daily and often at the beginning and/or end of each shift. In these meetings, sales managers and production control representatives spend a great deal of time telling the production people how badly they are doing, what they need to do to improve and what orders need to be pushed out quickly. Seldom do you hear reports on the health of the process and what needs to be done to improve **it**. If the process were looked at properly, there wouldn't be a need to shift managers around and have destructive or meaningless meetings.

When times get bad and the competitive edge is lost, how often are the future and present process needs of the company taken under consideration when a company is downsizing? How often is any consideration given to the resource and process needs of the company before or after the layoff? In this day and age, very little or no consideration is given to process health. Regardless of the damage caused to the process by downsizing, those remaining will have to do their best to keep the boat afloat. The business may have lost the people best suited to keep things together, but that's tough potatoes. It is now the task of the survivors to keep the widgets moving. Making sure that your process is in good shape after the downsizing is hardly ever the main consideration. Just cut the numbers and let the rest take care of itself. But what happens when business picks up again? But then, I repeat myself ...

Even the most people oriented companies have failed to give consideration to process excellence when downsizing. IBM was one of those companies that truly respected the dignity of the individuals who worked for them, or at least they did when I worked for them. The decision by IBM to downsize their manufacturing operations was a good one when it happened around 1993. I admire the way that they went about the downsizing where it relates to how the employees who were released were treated, and the manner in which the downsizing was carried out. There were some hard decisions that had to be carried out and the company did their best to complete the downsizing in a humane yet practical manner. It cost them a lot of money, but I believe that it instilled a strong loyalty for IBM into those that remained.

Still, having praised them for their treatment of the employees that were released, I still maintain that they failed to consider the health of their manufacturing processes when they made personnel cuts, and as a result lost more of their production capability than they cared to lose. I do not fault executive management, but I do think that senior plant managers could have done a better job of looking after their individual businesses. They carried out their cuts with priority on meeting the new resource numbers, not looking after their processes and the ability of those processes to function productively, albeit at a reduced level. Bottom line: They planned the whole thing well at the corporate level and implemented it poorly at the plant level. They were sensitive to the plight of those that would be separated, but they just did not give their processes enough consideration.

Ah, but there were those that gave a great deal of consideration to the health of the process. There were (and still are) those that believe that a quality process should be the primary consideration. They believe that you always work to improve your ability to be a quality oriented world-class producer regardless of your size. When in a growth pattern, make sure that a great deal of attention is given to where new machines are placed. Make sure that when a decision is made to hire new people, there is a demonstrated need for them. You hire or you buy shortly before you have maximized your capability, not just on a perceived need. If, on the other hand, there is a demonstrated need to downsize, it is important to understand that a quality process with the right people remaining on board is just as important after the downsizing as it was prior to the downsizing.

Trust Me – There Is a Better Way: Values Based Management (VBM)

I have spent a good portion of my career studying and emulating the achievements of the masters I studied. These were people that truly believed in the achievement of process excellence. These were people who believed what they preached, and acted out those beliefs. Fortunately, there are still those that believe that the best that can be done is to make a product as easy to produce as is humanly possible. They believe that it is wrong to ask for more output and then shackle your workforce with a bad process that makes it difficult for them to respond.

In my early manufacturing days, before I discovered the real truth, I did not give a great deal of consideration to my own process. I was interested in being seen as a good manager. I did what I did because I wanted to be a successful manufacturing manager. That meant that my boss also had to believe that I was valuable. I did not have a strong set of process values to guide me. I made decisions that were designed (I thought) to give me the highest probability of **being seen** as a successful manager.

Back then, if I were told to carry out an order that I thought would damage my department's ability to produce, or would threaten quality, I might whimper a little, but I would make the change. I'm not talking about decisions made at the CEO level, but those that emanated from middle and upper plant management. My management training taught me that all managers supported the decisions that were handed down. I later came to understand that it was the responsibility of each and every manager to enlighten his or her boss when they were ordered to carry out a change that was, in their opinion, not in the best interest of the business.

We owe it to the people to whom we report to explain that it is not a good decision to do what they are being asked when we know that it is wrong and we know that it would be a mistake to follow orders. Our bosses don't want to do something that would hurt their business, so if they make a mistake in judgment, it is our responsibility to at least offer them valid alternatives. We should make a business case for what we want to propose, not just tell them that they are wrong. We should convince them that we are right, not just tell them that they are wrong. If you are turned down, back up; review your argument(s) and present more convincing arguments to make your case. All levels of management want to do the right thing. They need experienced people to advise them properly and give them alternatives backed up by data and logic.

Many managers agree to carry out questionable orders and then telegraph the fact that they disagree with the decision to their own subordinates. They may not actually **say** that they disagree, but they certainly make their feelings known through body language or the tone with which they deliver the orders. I describe this sort of behavior as being **transparent**. When a supervisor goes out to pass on an unpopular order to his or her workers and is openly unsupportive of the decision through body language or tone of voice, the workers see that the supervisor does not support the order and that the source of the order is at a higher level than the supervisor. In other words, the supervisor is nothing more than a voice repeating the orders of someone they report to. If the order is unpopular with the workers, the workers know the perpetrator of the order and the supervisor is just a messenger, not a supporter of the order. Transparency damages the credibility of management.

Transparency is not only destructive in the workplace in a normal mode; it is especially destructive when change agents are out there trying to implement a new system. In many of the process improvement projects in which I have been involved, there were those that did not agree with the changes being made and were vocal in their disagreement. Now, it is one thing to argue against the changes with management, and it is quite another to argue your point in front of the employees who report to you.

In Part 8, when I talk about behavior modification and cultural analysis, I will explain how important it is to work with supervisors, managers, and technical leads in a positive way to promote healthy behavioral patterns and how adding this element to a process improvement effort greatly decreases the possibility of transparent disagreement. Gerry Clarke, now the president of Human Synergistics was the person that enlightened me in this area.

OK I Admit It – I Wasn't Always Totally Enlightened

When I first became a production manager, I truly did not know a thing about productivity and how to achieve it. I was given management training, but I was not given a course in how to achieve process excellence. I didn't even know such a thing existed. All of the managers, including me, just tried to make sure that our areas were not the plant bottleneck. To make sure that we were not seen in this light, we all tried to keep plenty of people and machines in place to meet the demand and the budget that we needed to support our needs. We didn't worry about the cycle time of our product, and we didn't worry about such things as proper production values; we worried about numbers, and what we needed to make the numbers.

Without guidance and training, we just did what we thought we had to do to get the job done. Looking back, I see that without a system and a set of good production values, nobody can step into a production management role and expect to do the right thing consistently. Buzzwords, gimmicks and pushy management are not the solutions that make us better producers. We have to have a system that provides us with a solid set of core values and a set of tools that brings discipline and repeatability to our process and the decisions that we make on a daily basis.

Eventually, through study, trial and error and a great deal of luck, I learned that I was seriously misguided. I prefer the word misguided to stupid. I learned this truth by studying the accomplishments and teachings of a very selective group of individuals who maintained that process waste is the true demon in manufacturing. Managing the financial end of a business is certainly important, but process stability and productivity is essential also. The production end of the business pays the bills. Everyone outside of production is there to support it. Management does its best work when it makes sure that it's decisions support the production effort. If you hurt production, you hurt the business. When production, management and support are united and work towards common goals, the business prospers.

These pioneers of process excellence that I'm talking about got a lot of attention by being successful with their own businesses. They proved the worth of their theories by improving processes to a point that other companies couldn't effectively compete with them. They kicked their competition's butts and did so by being the best in the business – the best of breed, so to speak. Champions in any arena have a way of attracting attention to themselves and their methodology. It is also true that great champions seem to appear out of nowhere, suddenly attracting attention to themselves by defeating impressive opponents. In truth, they probably worked very hard for a significant amount of time before becoming the best of breed. By the time they attracted attention, it was too late for their competitors to respond effectively.

There were many of us who took notice of the accomplishments of this group of pioneers and decided to find out more of what they were all about. I was one of those that came to believe in what these true pioneers were professing. Through their documented accomplishments, they helped to create the strong foundation that Fast Cycle Production is built on. I worked to build on their accomplishments, not outshine them. I learned from them that basing all of my decisions on a proper and consistent set of production values was the true and proper way to achieve process excellence.

Chapter 2 - Learning from the pioneers

In creating a methodology for process improvement, I turned to the masters: Deming, Taiichi Ohno, Shingeo Shingo, Hiroyuki Hirano and Eli Goldratt to name a few. Of course, some of their work had been going on for some time before I even became interested in process excellence or even simple process improvement. While I was out there trying to stay alive as a manager, they were out there finding new and better ways of approaching the business of production. They were trying to become world-class producers primarily through the improvement of their processes. They decided to be innovative and creative rather than accept the status quo.

Very little of the material that I teach today can be called new. I was reminded of this every time I worked with a new company. During the training classes, people would tell me that what I was teaching was not new to them. They would say that they've been there, done that, and may even be doing some of it now. **They would say that it worked for a while until they stopped doing it.** Some of them would tell me that they abandoned the "old" system that brought them their success. When I asked why they would abandon something that made them successful, they never came up with a good answer. In fact, not one of them could even explain **why** it happened.

They abandoned a system that brought them success and didn't really know why! How can this happen with rational, intelligent well-meaning people? Although there are no definitive answers for this seemingly irresponsible behavior, I found that it happens more often than not, especially with large companies. The answer to this enigma seems to be associated with management. The management that was in place when the improvements were made was often not the same managerial team in place when the reversal of productivity came. We are not talking about a complete turnover, but rather, just a few key managers that caused the deterioration of production values.

But executive management could have kept the productive system in place if they had **constantly** made it clear to their reports that they supported the new system. Instead, they allowed their productive gains to be lost through neglect. In a few cases, the plant manager, after being successful in improving their productivity was promoted, or left the company. A new person that just didn't understand or subscribe to the new production values replaced them and the downward slide began. After all, if the new person truly wanted a system to be maintained, and they made that intent clear, it would definitely be maintained. The lack of managerial focus and resolve was what allowed the destructive changes to happen. Destruction through neglect – what a waste!

Don't Blame The System ...

Many of the people who helped to form my beliefs did so long before many of you were born or even of college age. Forming the beliefs of my system took many years, but it was time well spent because it allowed me to put together a system that can be implemented quickly and still produce dramatic results. There was a lot of trial and error in the formation of this system, and in the beginning there were a few outright failures.

My first experience with **Kanbans** was a disaster that I will never forget. Kanbans as a tool did not fail; it was the way that I tried to use them that failed. My first attempt at teaching a group of people how to flowchart their process was another downer. Again, flowcharting was not the failure; it was the way that I tried to do it. Yes, I got bounced around a bit, but I managed to pick myself up, dust myself off and learn from both my successes and my failures. **Now I have a system that never fails and always produces dramatic results.** The system works because the values that form the system are rock solid.

Many companies have tried to make JIT work but failed to reach the productivity levels that they aspired to. However, you can't put the blame on JIT. In their attempt to become more productive, companies will try just about anything and will just as frivolously abandon whatever they start if it does not work instantly. Whatever their reasons for trying to become better producers, the fact that they failed at some of their attempts is not necessarily a condemnation of **what** they did or **what** they tried. It may be that they just did not implement JIT correctly.

They may have had the right idea and even the right system, but just did not know what they were doing or how to go about it. It could also be that they really did not believe in what they were doing and went about it half-heartedly. The most likely possibility is that there were some that really believed in what they were doing, but they failed to get upper management support. Still, even if they made an attempt to improve and failed, it would be important for them to understand what they learned from that failure.

In my case, if I were to deduce that Kanbans were useless in **any** process because they failed on my first attempt to use them, I would have made a very serious error. Instead, the lesson that I learned was that I did not implement them properly. If I had given up on setup reduction the first time that I failed to reduce the setup time, I would have missed many, many opportunities. Could it be that HOW you go about doing something is as important as WHAT you are doing? I now know that to be true!

I was able to overcome my failures by learning from them and applying my newfound knowledge towards improving my methodology and making sure, or at least attempting to make sure, that I did not make the same mistakes twice. Fortunately for me, there was an abundance of great people to guide my thinking. Failures only have a devastating effect when you fail to learn from them. But in the competitive environment that we have today, companies should also learn from the mistakes of others and use the systems and tools that get great results. It is possible to avoid the pitfalls that disastrously affect others. The goal of this book is to help others to avoid failure by giving them the benefit of my experiences.

Many of the people that influenced my philosophy are gone now, but their individual accomplishments live on, and I am proud to say even live on through FCP. My belief in them and their teachings remains unshakeable. Recently, there has been a resurgence of interest in the values and tools that came out of the Toyota revolution many years ago. The pioneers of the 60s, 70s and 80s that were able to bring Toyota from a second rate automobile manufacturer to world prominence are still being emulated today. Kaizen, Kanbans, Lean Manufacturing and The Five Why's (Root-cause analysis) are still being used today in one form or another. There are many consultants, or *senseis* out there teaching what was created decades ago. If they are embracing those old values, chances are they are not too far off track. The only problems they may create for themselves are the problems they experience from going about it in the wrong way with a flawed implementation methodology.

I think that it is important to point out that the masters of the Toyota model did not get hung up on buzzwords but instead, focused on substance. Today, everyone seems to be obsessed with being able to use the old Toyota buzzwords like *sensei, kaizen, kanban* and *muda.* It is even quite fashionable to be able to "define" each one of the words. Let me tell you that not many people **really** know the *true* meaning of those words and how, why and when they should be used. Translating a buzzword is one thing; truly understanding it is quite a different story. If I accomplish my objective with this book, readers will learn the true meaning of the terms and their proper application.

There are some people out there that have established themselves or have been recognized by others as *senseis*. There are also people out there who have achieved the title of doctor. We all know that there are some very good doctors out there that are a model of their profession. There are also some doctors out there that stink up the place. The same is true of *senseis*. Having the depth of knowledge to be able to transfer knowledge to your students is part of the description of *sensei*. The other part of being a *sensei* is to provide leadership and always to do what you would teach others to do.

Speaking of buzzwords, I remember many different "quality programs" in the 80s, each of them having some temporary success, only to eventually fall out of vogue. The company that I worked for tried out most of them with varying degrees of success. I can remember getting very excited about **Quality Circles**, only to watch it disappear for lack of interest and support. For a while Quality Circles worked great for us because it really got the employees involved in the decision process. We had meetings twice and sometimes three times a week and we got some great input from the members and even made some positive changes. At first, we were empowered, then all of a sudden, managers started complaining because their best people were tied up in meetings "all of the time." Pretty soon, people failed to show up for the meetings because, "they were needed in their area." Eventually, the meetings just kind of disappeared.

I also remember being the manager that was given the responsibility for establishing our **Employee Suggestion Program**. Here was another great employee involvement thing. What a concept: Get the employees to make suggestions that would make things better. We thought it would be great for our company, only to find that we created something useful and then ruined it by being too critical of other suggestions. We established a panel of experts that would evaluate each suggestion and give our approval to the ones that the experts liked the best. It was not too long before it was almost impossible to get a suggestion accepted. The suggestion would have to be the equivalent of being able to make gold out of lead.

I made the fatal mistake of implementing a potentially good, long-term suggestion program and then making a short-term program out of it. We came out of the blocks really well, but we lost ground very quickly. I failed to see that the really good suggestions that proposed doing away with useless and meaningless tasks were labeled as gripes instead of being worthy of consideration as they should have been. "We need the redundant counts and inspections," we said. "It is important that you fill out that meaningless paperwork. How dare you try to get rid of something so important to our customers?" I'll bet that you're surprised that I could be so misguided? Well let me tell you that I led the pack of misguided fools for some time. I took stupid pills by the handful.

I remember reading a story about a Japanese company that started a new factory in the U.S. One of the first things they did was to instruct their new plant manager to start a suggestion program. He did so and the program was successful. So successful, in fact, that he was given an award by upper management. A year later, he was fired!

Although the suggestions were pouring in, he was not **using** any of the suggestions, or the number of suggestions that he used was very small. He had made the same mistake I made when I started a suggestion program: If you do not use their suggestions, or if you make it very difficult to get one approved, the suggestions will slow down and eventually stop. The message here is that you cannot ask for involvement unless you mean it. You cannot say that you want someone's help and then refuse to use it.

I remember teaching my people about the difference between **Quality Control** and **Quality Assurance**. I taught them that the former was trying to achieve quality through the inspection of the product, which was bad, and the latter worked to achieve quality through the inspection of the process, which was good. After teaching them the difference and telling them that Quality Assurance was most certainly the right way to go, I went out and had them inspect the product. I just couldn't take the chance that my department would ship bad product to our customers. When it came to quality, you just can't take chances, right? So you **compromise** your values and you justify the action by stating that you are concerned about your customers. My point here is that I **told** them to do one thing, and I did the opposite. I was **not** prepared to, "talk the talk **and** walk the walk". I was simply **all** talk.

In working with a number of companies, I have found that in many cases, Quality Control (inspection of the product) is vitally important. There are some companies with products so complex and sophisticated that product inspection is not only necessary to ensure product quality, it is necessary in order to keep customers. Without inspection, defects that could be caught prior to shipment would reach the customer. Quality assurance (inspecting the process instead of the product) is very effective with product that is not highly technical. In such cases, inspection of the products is not really necessary. But in some companies, it would be impossible to achieve the level of quality assurance necessary to ensure the near perfect quality necessary to satisfy customer demands. But with either quality assurance, quality control or both, the really important fact is that quality should be more than just a program.

Quality was more than a program to the Toyota masters. Another story that came out of the Toyota revolution was about an American industrialist that was visiting Japan to hopefully learn from them. He asked his Japanese counterparts what the name of their quality program was. His hosts replied simply that they did not have a quality program; quality was a way of life with them. Quality as a way of life – what a concept, eh? We live in a world where quality is nothing more than a department.

Another *Quality* story that I like to tell is about the VCR business, and although I believe the story to be true, I cannot say with certainty that every detail of it is accurate. It exemplifies my theory of what quality is all about, so I use the story in my classes. It illustrates how quality and process excellence reached new competitive highs.

I heard the story at a quality seminar for the electronics industry. The speaker was glorifying the Japanese practice of creating vendor families and was talking about component testing by the vendor, as opposed to testing vendor supplied parts after they were received from the vendor. As the story goes, even though oriental companies eventually dominated the VCR market at the time, it was once primarily an American product. In the beginning, the American companies did OK, but they lost market share over time and were eventually forced out of the business. Eventually, only the Japanese continued to make VCRs. How could this happen? Powerful American companies couldn't compete with the Japanese when it came to quality and price.

"Who do you think," the speaker asked the audience, "does the functional testing of VCRs in a Japanese factory?" After hesitating for a few moments while scanning the blank stares of the audience, the speaker triumphantly says, "You (the customer) do!" He went on to explain that there is no functional test at the factory because the process is so repeatable and the qualities of the components are so predictably good, that there is no need for a functional test! The functional test is completed when the customer turns on the VCR for the first time. The cost that goes into staffing and maintaining a functional test department is saved. They do without a functional test department and still their quality improves.

I love to tell that story because it epitomizes my belief that you cannot just eliminate inspection, nor can you just eliminate functional test or anything else that does not add value. You must find a way to eliminate **THE NEED** for them. You can't just order something's demise. Nor can you shout MUDA (Waste) three times in rapid succession and expect it to disappear. You simply find a better way, and then do what is right. The bad is substituted by the good.

NOTHING that slows you down in production can be eliminated by simply tagging it as waste. If you want something to go away in the production world, you must find an alternative for it. You must find a solution to the problem! Identifying a process step as waste is a good beginning, but it is only the beginning. It is not so simple that you can say that you have found waste and the next step is to find a way to eliminate it. To do so, you have to understand, if you can, the original reason for that process step. Finding the original reason gives you insight into why that step was added to the process. Sometimes you try to learn the original reason and cannot do so. Nobody knows why it exists. It may have been there long before anyone there presently was hired.

Is It A Buzzword, Or Do You Really Believe?

I teach that quality can never be a program or a buzzword. You have to believe in the importance of quality *so much* that every decision you make, every change you implement and every solution you come up with, cannot possibly cause the quality of your process to suffer. Your concern for quality has to be indelibly built into your decision process.

Poka-Yoke (mistake proofing) is another term that should not be bandied around lightly. If you understand it's meaning, you must still believe in the concept. By translating the word into English, you have not proven that you are an advocate of its use. *Poka-yoke* is incorporated or inserted into the system. Shingo would say that a checklist could be *poka-yoke* if that checklist were created to improve quality, not just add a useless redundant step to the process. He *would not,* in my opinion, say that a redundant inspection was *poka-yoke*. A redundant inspection would be waste!

In my opinion, Shingo had a big problem with sampling inspections. He acknowledged that the reason for them was the high cost of 100 percent inspections. He maintained that sample inspections just did not work statistically. You may catch some defects, but what were you missing? In his search for a practical way to achieve total quality, he introduced an alternative to sample inspections, rather than denounce them statistically as impractical. He went for an alternative, and then proved that his system worked.

While 100% inspections are more significant than sample inspections from a statistical standpoint (but not from an economic standpoint), **successive checks have been found to be more significant than 100% inspections,** and they **are** economically feasible. The successive checks are another example of *poka-yoke*. I have applied the spirit of *poka-yoke* with my application of **single-unit flow** (Part 2, Chapter 2). I promote the use of single unit flow for complex inspections, where a team of inspectors each inspect portions of the product individually, and then double-check what has been inspected prior to their own inspection. I have promoted this concept to clients as being superior to a group of inspectors individually inspecting the complete product and then inspecting it again later. My point here is that I tried to use the essence of poka-yoke. I found that single-unit flow not only results in better quality, it allows for increased output with fewer resources. Properly applied, it is superior to normal detailed inspection methods.

In the late 1980's, Hiroyuki Hirano came out with a book that turned out to be a brilliant introduction to the JIT production system. Mr. Hirano is an eminent expert in the JIT production system and has written several excellent books on how to implement JIT. Most recently, he wrote another book, "5 Pillars of the Visual Workplace: The Sourcebook for 5S Implementation". In my opinion, Mr. Hirano successfully points out in his book the importance of **preparing** a company for any type of productivity improvement by first focusing on the 5S's: Organization, Orderliness, Cleanliness, Standardized Cleanup and Discipline. Of course, Mr. Hirano is referring to the Japanese "S" words: Seiri, Seiton, Seiso, Seiketsu, Shitsuke. Having said that, I still maintain that Values Based Management decisions not only provide the same benefits, they do so in a continuous fashion. Values Based Management, along with properly applied cultural enhancement tools (Part 8, Chapters 1 and 2) provides a company with an effective long-term solution.

Again, I want to point out the importance of understanding the true meaning of what the masters of JIT have been trying to promote. They are not, in my opinion, promoting buzzwords, but concepts and values. They are pointing out to all of us that **the way** that we approach and implement process improvements is extremely important to the probability of success both short and long term. They are trying to teach us that it is not enough to just implement their tools and concepts, but to SUSTAIN them. **Too often I have seen companies implement with exuberance, only to later abandon with indifference.**

Process Analysis and the Elimination of Waste

I will cover this subject in more detail later, but first I want to explain why my method of flowcharting and process analysis differs from standard flowcharting. When I was working to implement CFM (Continuous Flow Manufacturing) in companies, process analysis was the first thing that we did (Part 2, Chapter 3). We trained people in the traditional definitions of Value-Add and Non Value-Add. A Value Add (VA) process step is one that adds value to the product from a **customer** perspective. Adding wheels to a car adds value because the customer does not want a car without wheels; therefore, the process step(s) that include adding the wheels are Value-Add steps.

On the other hand, a Non Value-Add process step is defined as one that **does not** add value to the product from a **customer** perspective. Inspections, tests, paperwork and engineering actions are steps that do not add value to the product. The customer does not care how much you inspect something; they just want a quality product. If inspections were value-add, then adding more inspections would make the product cost more to the end customer and the customer would feel good about the extra cost.

That being the case, the old types of process analysis stigmatized those areas and jobs that were deemed **Non Value-Add** even though the action taken was seriously important to your product. There were those that resented the notion that their job added no value to the product. Who likes to be told that their job is non value-add? For many years I went along with the standard definitions of value-add and non value-add, but eventually came to understand that there had to be a better way of analyzing a process and taking out the waste without approaching certain areas in a threatening and destructive manner. People want to believe that what they do is meaningful and important. When they get the notion that what they do adds absolutely no value to the product, what do you think their internal response will be? Of course, they will be resentful.

What a warm feeling to hear that what they were hired to do is nothing more than absolute waste! I say this as a joke, but that is just what we do when we (an improvement team) call something waste. We just condemned someone's job. When I re-designed process mapping and analysis for Fast Cycle Production, I did so by evaluating each department separately. I did this because I knew that there is waste in every department regardless of what it does. Furthermore, when you talk about defining value from a customer perspective, I believe that you are talking about your internal customer, not the external customer to whom you sell your product. **For the reasons mentioned, it is extremely important to never define a department as VA or NVA. Only process steps are VA or NVA.**

Each department has a **product** (That which they were hired to produce). If you work in an assembly area, your product is assembling components and/or sub-components to something else. Assembling components is Value-Add. If you work in an inspection department, your product is inspection. That is what you do! You make your customer happy by doing your inspections well. That is your Value Add. Anything else that you do, such as paperwork, rework and even redundant inspections are Non Value-Add.

This kind of analysis is what I teach and I believe it to be the correct way to approach a process mapping and analysis endeavor. Searching for the demons in your process is one thing. Singling out whole departments and workers as being part of the waste is another. I have found this system to be very useful when I worked with non-manufacturing types of businesses such as, loan companies, banks, homebuilders and some service organizations. Of course, it also works very well in manufacturing. It is the correct way to do value analysis and to identify waste.

More Values From The Masters

You understand that the literal translation of *Muda* is waste, and you know that waste is something that you do not want in your process. Nevertheless, you go right out there and make decisions that insert waste into your process. The JIT and CFM experts taught that you should never have more Work In Process (WIP) on the floor than you need. I have seen company after company dump excess WIP onto the production floor with the absolute belief that they were doing the right thing. They had this belief that "If you dump it, they (the uninformed masses) will make it move". Even though their process can only handle so much and no more, these titans of industry continue to dump the product into production and watch it pile up somewhere on the line. Then they go to the people on the production floor and ask why they (the people they just handcuffed) cannot produce as much as they should. This goes **beyond** stupid. To act that dumb, you need chemical help.

It has always been mind boggling to me that this kind of mentality exists among normally intelligent people. They actually believe in their heart of hearts that dumping the excess work onto the floor will actually improve the ability of the process to handle the work when, in fact, they are creating more work that adds absolutely no value to the product. They create more work for their people by forcing them to **manage** the WIP. They also know that when WIP sits around, bad

23

things are likely to happen to it, but they do not just stop there; their next step is to create HOT LISTS. I will elaborate on this point more when I cover **Modern Manufacturing Values,** but for now my point is that people who pretend to understand the meaning of the word **Waste**, actually go out there and create it! Something is wrong with that picture.

The NEW Enlightened Tom

Did I mention that I was once one of those mental giants? Yes, your honor, I led the pack! I would drag more work into my department than it could ever do in a month and feel darned good about it. I had this warm and fuzzy feeling inside when I saw all that work on the line. Because my people respected and trusted me (?), they also came to believe that WIP was wonderful. If they got down to a million or so widgets, they would become uncomfortable. As a leader I sure led them astray. It sure was confusing to them when the new, enlightened Tom came forward and told them that what I had been teaching them all along was just bull hockey. They recovered when I reminded them that I was a member of management.

Fortunately, I saw the light and went out to teach others not to make the mistakes that I had made. However I found that just teaching people that it was bad to do things like that, is not always enough to stop them from doing it. I found that when some of the workers were under pressure to get product out the door, they would go right back to the old practice of dumping work on the floor. A customer puts pressure on them and they waffle and release the product knowing full well it is going to bog down somewhere on the production line. Their only logic is that they are doing what the customer wants. Of course if the customer were convinced that the production control manager should commit suicide, they would do that also, right? I don't think so!

Today, I still struggle with this tendency of some of my clients to waffle because I know that my intention was to create true believers and I have failed in their case. I have failed to instill the **value** in them that it is counter productive to do that sort of thing. It is as immoral as stealing because you **really** need the money, or cheating on your taxes because everyone else is doing it. Violating Kanbans because you want to "help" the customer is as productive as shooting your foot to ease the pain somewhere else. To some, Kanbans are places where WIP accumulates. What does it matter if we "increase" it a little? It will all work out. Besides, Kanban is a buzzword to some, not a usable tool. Kanbans and Dynabans will be covered in much more detail later (Part 6, Chapters 1 and 2).

My first "Sensei" was Dr. W. Edwards Deming. Although he never knew my name, you could say that he mentored me from afar. Actually, I read his books (several times) and became a true believer in the Deming philosophy. Deming's 14 Points for Management not only affected my thinking but also permanently altered my management style. His analytical approach to finding the root cause of the problems experienced in manufacturing was an inspiration to me and was instrumental in the creation of **TAKT Management** and **FCP Value Add Analysis**.

I don't know if everyone got the same thing out of what Deming taught, but I can say what I personally read into his words. I feel that Deming believed that managers should be leaders. Being competent managers is not enough. The word "leadership" is mentioned in his 14 points a number of times. He encouraged managers to eliminate work standards, management by objective, management by numbers and numerical goals. He strongly encouraged cross-functional cooperation and he insisted that quality couldn't be achieved through inspection, but could be achieved by building quality into the line.

It **is essential** for managers to provide proper leadership to their employees. I use the word "proper" because in my opinion, a good leader could have bad production values. My experience with many manufacturing managers, and my experience with the old, unenlightened Tom has led me to believe that leadership is a word that is often used, but is seldom given the priority and consideration that it deserves. It is like saying that you believe in world peace and then you go out and start a war. Managers must show leadership and do so on a daily basis. They cannot be Jeckyl and Hyde types. They must be consistent and repeatable. Most important of all is that they be good leaders with solid production values.

At a business seminar that I attended, a gentleman there shared with the audience his personal experience with W. Edwards Deming. Deming became famous for his work with the Japanese and his published work is still actively selling today. To the Japanese, he was a genius who helped them to become world-class producers. Prior to his success with the Japanese, it was said that he had been totally ignored by American industrialists. The Japanese were quick to embrace his philosophies - almost religiously. To them, he was a genius. They were right!

The gentleman at the seminar told the story about the time that he attended one of Deming's seminars. Of course, this was some time after Dr. Deming became famous. At a certain break period in the seminar, Deming would autograph personal copies of his book for the attendees. Personally autographed books by Deming were highly prized acquisitions in those days. The attendees would stand in line and eagerly hand the master their books to be autographed. Most of the time, it was said, Deming virtually ignored the person before him. Occasionally, he would look up to acknowledge the existence of the person standing before him.

While Deming was autographing this gentleman's book, he happened to look up and asked the gentleman what his name was and where he worked. It should be noted for those of you that have not attended a Deming seminar that personal recognition from Deming would be considered an honor to anyone and was therefore certainly welcomed.

The man felt extremely lucky and honored to be acknowledged by this great man and proudly answered that he worked for a very large international company. (The name of this company shall remain nameless for reasons that will soon be obvious) Well, as soon as he heard the name of the company, Deming's face lit up and he asked the man if he could talk to him at the end of the session. Of course, the man was honored and eagerly agreed to a meeting.

The time and place of the meeting were agreed upon. The man eagerly anticipated his private session with such a brilliant man so much so that the thought of personal time with Deming dominated his thoughts for the remainder of the session. He knew that he would remember this session for the rest of his life. He worried that Deming would forget all about it and his bubble would burst.

He was wrong. Deming actually sought him out after the session and seemed eager to talk to him. After confirming that the man did indeed work for that same large international company, Deming said, "Your Company has the best managers in the world!" Well, the man flushed and started to thank Deming for the compliment when

Deming interrupted him by saying, "But your company has no **LEADERS**."

Dr. Deming was not attempting to be insulting. He was making a point that outstanding managers were not as valuable to a company as they would be if they were also outstanding leaders. Leadership to Deming was as important to companies as it is to nations.

Did you ever hear of "World Managers"? ... Of course not!

"Outstanding Management" as a compliment or reward? ... Forget it!

Outstanding LEADERSHIP is the true compliment!

I often define "leadership" as the ability to convince people to do something that they would not normally do because you are asking them to do what you believe is right. Having said this, it should be clear that if you are looking to be a change agent by implementing a new production or business system in your plant, factory or business, you need leadership ability. Being a top manager in your company does not by itself fill the prerequisite. True leaders can influence the intended path of their followers. Their paradigm and even their instincts may lead them in one direction, but they succumb to leadership because they believe in the leader. If you really think about it, most people follow their boss's **orders**, not their **leadership**. Leaders always believe that what they want their followers to do is truly the best thing for them and the company.

Think about it, when you are listening to true leaders, you may have some questions or concerns in your heart, but you will go with their recommendations or requests, because you believe in them. Let me reiterate that point ... **you believe in them!!** Whether it is war or work, true leaders do well for their followers with respect to what is right. I think that in many cases, workers can make even the near impossible happen because of their strong belief in their leader. They believe so much in them, that they make the leaders wishes happen. Leadership is a powerful motivator of people.

I have worked with dozens of individuals that have been charged with the task of implementing positive change. There were some that were suited for the task and there were some that probably should not have been on the team. Being good at your job, or being between jobs and available does not hack it either. When you put together a team of leaders, properly motivated and trained, and armed with the right set of process values, just give them the baton and watch out! You can bet your bottom line that they will produce for you big time!

Deming was a pioneer who had a huge influence on the development of my management philosophy. I learned much from his analytical approach to problem solving. Through him I learned that root cause analysis is the correct approach to solving problems. Although I did get into Pareto Charts for some time, the application of Deming's teachings came to fruition in TAKT Management and my approach to simulation modeling. Deming was truly a brilliant man and a pioneer in his field. Fast Cycle Production philosophy and methodology has much of Deming's teachings contained within it.

FCP is a continuous improvement process with the emphasis on **process improvement**. Not only does FCP yield dramatic cycle time and operational cost reductions, but it **always** produces an improvement in product quality. As Deming teaches us, we recognize success when, and only when, we delight our customers with the product they want, when they want it, and at the quality levels that they expect. All of these conditions must be met before we declare any productivity improvement project a success.

The man made a believer of me. To this day I still teach what the master taught me (in absentia). Eliminate *the need* for anything hurting you. Do not just eliminate process steps without finding alternatives to the wasteful steps. Try to find the reason that the waste occurs. Something made it happen – find it! Provide proper leadership to those that I teach and always remain true to my values. Lead through example! Deming started it all, but there were others that altered my thinking and helped to create my productivity improvement system.

Another pair of pioneers in the area of process improvement was Taiichi Ohno and Shingeo Shingo. They were the architects of the Toyota Production System that kicked some American butt for a long time. Detroit just did not have an answer to this pair. They brought Kaizen, Kanbans and Setup Reduction to the table and served up some serious servings of continuous improvement. Most importantly, they embedded the belief in me that VALUES are important in the work place. It is not just a work place where "doing anything necessary to get the job done" is appropriate. They taught that you could have your cake and eat it too.

Shingo's work with setup reduction is legendary. He wrote a book about setup reduction, "S.M.E.D. Single Minute Exchange of Dies", that I still consider the bible of setup reduction. In his book Shingo talks about how many companies try to offset the effects of setups by simply increasing batch sizes. The logic of this belief stems from the fact that, if you avoid setups by running a large number of parts between setups, you will increase the amount of time you spend producing those parts and minimizes the amount of time required for setups. Very good logic when looked at singularly.

Their logic rings true and works when you only have a limited number of part types. Why do a setup if you do not have to because setups decrease your output potential? On the other hand, if you have a large number of part types, how can you run one part type all day when you have other part numbers to get out? All you accomplish is running up the amount of WIP on the floor.

Shingo insisted that increasing batch sizes simply is not the way to go. If you just increase the batch size in order to increase your output, you are working with the wrong variable. Shingo said that you should work to **decrease** your batch sizes, but in order to do that you must first reduce your setup time dramatically. Setup time is the variable you should be working on. Setup reduction is now an important element of Fast Cycle Production and is covered completely later in the book (Part 5, Chapters 1 and 2). Shingo proved that setups could be reduced from days to hours, and hours to minutes, and even to one minute. That is why he named his book, "Single Minute Exchange of Dies".[1]

Consider the example of an automobile assembly plant back in the early days when there were maybe three or four colors to deal with and deliveries were made monthly. With that few colors to deal with, one could easily paint nothing but black cars for a week, white ones for another week and so on. At the end of the month, the "color"

---

1. *A Revolution in Manufacturing: The SMED System.* Japan Management Association, 1983 (English edition Productivity Press, 1985)

requirement will have been met. There was no need for doing time consuming setups every few cars or so. Batch painting was feasible because you did not have many colors to worry about.

Enter today's world where daily you are required to ship product in a variety of colors. If your setup took four or five hours, you would not be able to make your output requirements. You would be spending too much time doing setups. Fortunately for the auto industry, it is now possible to change from one color to another in less time than it took you to read this paragraph. The setup takes less than a minute and that includes cleaning the paint hose. Such a reduction in setup time would have been unheard of several decades ago.

The beauty of Shingo's system is the simplicity of it. It is not only easy to understand conceptually, it is a solution that is simple and easy to implement. Like most of the JIT tools, it presents a repeatable process. Although I have made changes to the way I implement setup reduction on the floor, the basics are still Shingo's. I will elaborate on this in Part 8, but for now suffice it to say that instead of the engineers having sole responsibility for setup reduction, I have made it a team effort, one that includes the people that have to complete the setup – the workers. FCP setup reduction includes quality, maintenance, engineering and workers in its setup reduction methodology.

Taiichi Ohno, who many consider the main architect of the Toyota Production System, eventually became a senior manager at Toyota and is still thought of by many as the man that earned the bacon for Toyota. He created the Kanban system of WIP control that is also brilliant in its simplicity of operation and implementation. The Kanban Control System is still used by many companies to control WIP bubbles and keep cycle time low while maximizing output. Kanbans are still widely misunderstood and misused by many companies today.

I teach Kanban control because in my opinion, Kanbans are excellent tools for smoothing WIP throughout a factory and I know of nothing that can do it better (Part 6, Chapters 1 and 2). I have modernized Kanbans to some degree but as I will point out later, the basic tenets remain. It is very difficult to improve upon something as brilliant as the Kanban, and my change was only a change that responded to a problem that exists with a large product group variety – Dynabans (Part 6, Chapter 2).

These two Japanese pioneers and others taught me that values are so terribly important in production. You can't just put rules out, and expect people to follow those rules just because you tell them to. You have to show them that you live by the values you profess. It is no different in the raising of your children. You cannot just say to them that it is wrong to do this and that and then go out and do the opposite yourself. They might follow the wrong path if you fail to set the proper example. You do not want them to do something that you know is wrong, just because they saw you do it. What kind of leadership would that be?

You work to develop the right production values in your employees, as you would work to develop the proper personal values in your children. You know that you won't always be able to be right there to guide your children and help them make the right choices, so you work to instill values in them that will guide them when they have to make the tough choices. The same is true with your employees. If you have done your job correctly and set the proper example and leadership, your employees will take the right path.

The decisions that you **and** your employees make will be good ones because those decisions will be based on sound values. They won't give in to pressure either, because they believe strongly that what they are doing is right. For you to instill values in someone, you must provide them with leadership by being a model of what you profess. Talking one way and acting another just doesn't work. You can't read both sides of the page at the same time.

In a factory, where everyone from management to line worker have the same production values, the decisions that the manager would make in an unforeseen situation are the same as the decision that the line worker would come up with in the same situation. The people on the production floor have no need to hold up production until they can find their supervisor because they know what he or she would do. That process would run like a top regardless of the kind of problems that were experienced. This utopia that I speak of exists, but only where identical values permeate throughout the factory. **Given a certain situation, every single person on that floor would make the same decision because they all share the same value set.** What a place that would be, eh? Well, that is the essence of **Values Based Management** in Fast Cycle Production.

Constrictions, Bloody Constrictions

If there is anything that you can say is always true, it is that all work centers are not created equally. If they were, you certainly wouldn't need Kanbans to smooth the WIP on the floor and you wouldn't have to worry about a single workstation getting behind. No, the truth is that some departments process the product quickly while some take longer, sometimes much longer to process the same amount of product. They are not equal in ability because they simply do not have identical tasks or abilities. The work centers that take the most time to get their tasks completed are called bottlenecks or gates. The slowest of these is the constriction (pinch point) in the process. Constriction theory was given a lot of attention with JIT and CFM and it is also an important element of Fast Cycle Production (Part 4, Chapters 1 and 2).

It seems to me that production managers want to understand constriction theory, but fail to get the message when it gets down to application. It just seems that production management just cannot come to grips with the fact that it is important to use their resources to open constriction points. I believe that this anti-productivity logic exists because some people feel that the easiest way to deal with constrictions is not through analysis, but through pressure. It is amazing to me that pressure is almost always preferred to analysis. But there was one person that helped to change that thought process. That person was Eli Goldratt.

To me, Eli Goldratt is to constriction theory what Shingo was to setup reduction. He literally rewrote the book on constriction theory. In every engagement, I either encourage my clients to purchase Eli's book, "The Goal",[2] or, in many cases, I buy the book for them because I want them to understand the importance of what I call "Gate Busting". I did not learn constriction theory from Eli's words as much as I learned **how to deal with constrictions** and eliminate them.

Whenever two people read a book, they may get different messages from it. The message that I got from Goldratt and what I teach to others about constriction theory is that in the world of production, management never treats the bottleneck the way it should

---

[2] *The Goal: A Process Of Ongoing Improvement; Eliyahu M. Goldratt; North River Press; 08/1994*

be treated. In reality, the bottleneck work center gets the worst of treatment by its own management.

"The Goal", in my opinion, pointed out that we often fail to deal with the bottleneck properly. When the bottleneck can't produce enough widgets, we put pressures on it to do better. We threaten, intimidate and generally use only negative methods to get more out of it. We seldom give it what it really needs, and what it really needs is **HELP**! We think that we are just "encouraging" the bottleneck to do better. What we think of as encouragement often has the opposite effect on the bottleneck. What we think of as encouragement is really pressure.

Dealing with an area that is the bottleneck is always a challenge. First of all, the people working in it know full well that they are the only area not meeting requirements. They are reminded constantly of this fact so much that they are usually sick of hearing it. Presenting employees with work standards (quotas) or numerical goals certainly won't work. Work standards serve no purpose other than telling employees what the minimum standards are. They do not help to produce more. Pushing numbers on people has never worked effectively before, and will never work in the future no matter how "humanely" managers think that they apply this pressure. People working in the bottleneck area can tell you this. We have to stop telling them to start working harder or faster and **help them to be more productive.** There's a difference between helping and telling - a very distinct difference!

When someone who works in the bottleneck area goes to a production meeting, there are a lot of words spoken, and the translation to those words is most likely, "We would be doing OK if you (the bottleneck) were not dragging us down!" When you work in one of those departments, you simply do not want to go to a production meeting. If for some reason you do have to attend, you had better have some good "excuses" for not being able to ship as much as is expected.

It is quite normal for people who **do not** work in the bottleneck area to blame their own inequities on the bottleneck. They treat the present bottleneck area as a permanent entity. To them, the bottleneck will always be the bottleneck. Since the bottleneck will always be another department, they often fail to work on their own problems

33

because in their mind, there is really no reason to do so. They are already faster than the bottleneck so why bother to improve?

In a company practicing FCP, every department had better worry about their own production and not think of themselves as being safe from scrutiny. If there is no improvement, a previously safe work center may soon find that they are the bottleneck because everyone else is improving but them. With FCP, the people working in the bottleneck area know that help is always on the way because analysis is constant. There is a simple fact of life in production: Working in a bottleneck department or work center is not preferred.

Have you ever noticed that in companies that are having trouble meeting requirements, there are certain departments that nobody wants to work in? I'll wager that in most cases, no matter what type of company is involved, the area that nobody wants to work in is the bottleneck area. In many companies it is not career enhancing to work there – people who work there catch too much flak! They use every method possible to keep from working in the bottleneck area. Managers and supervisors consider a transfer to a bottleneck area a kiss of death. If possible, they will use their influence to stay out of it if they have any, or they will try to get transferred out once they find out what it is like to work there.

People don't understand that collectively their line cannot do any better than the bottleneck is able to do. The process cannot put out more widgets than the bottleneck is able to put out. To treat the gate people badly just does not make sense. What they need is help! If you really want to help the constriction, you can start by not insinuating that they are incompetent scumbags out to make everyone else fail. If you really want to improve your process, start treating the bottleneck area as family and start helping them through analysis.

Chapter 3 - Developing a new set of values

There has been a lot written about how to be a successful producer. Shingo, Deming, Goldratt, and others all thought that they had the answer. There can be no doubt that all of my predecessors believed that they had the perfect system. They all were able to brag about their gains, and obviously they accomplished a great deal. But becoming a successful producer is one thing; remaining one is something else.

Recently, some of the old Japanese Just In Time tools have had a resurgence of popularity. Because they were successful in their day, the thinking is that they will be successful now. They have been incorporated into new productivity improvement schemes, many of which have methodologies that are similar to FCP. The similarity is only apparent at first glance. FCP is superior, not because it is so different, but because it has the best approach to the achievement of true process productivity. It is the sum of the parts and how those parts are applied, not the individual collection of parts that makes Fast Cycle Production so effective.

But no matter how effective a production system is, to maintain the productive gains you must stick with it. Even with the best system in the world (FCP), you have to stick with it for the long haul or you will lose everything you have gained. Abandonment means defeat! Getting there is only half the battle – the continuity of the system in its purest form will provide benefits indefinitely. Compromising the values will cause erosion of the system and eventually extinction. This is true with any system implemented; including FCP. How did I discover this fact? The answer is that I lived it myself and I've seen the erosion with many companies. Many of my clients had seen both success AND failure.

Faster I say! Give me more!

When I was a misguided soul, I had a need for speed and numbers. Deming must have had me in mind as the "before" model when he made up his 14 points for management. Beating output expectations and doing so consistently were the **only** report cards I used. I pushed my own people to work faster and get more PCB's (printed circuit boards) out. My logic was simple: If they worked faster, they would get more product out. I had the idea that low cycle time meant working faster. Boy was I wrong! Today I still worry about cycle time, but I teach that reducing cycle time has nothing to do with working harder or faster. From Shingo I learned that the proper way of reducing cycle time was by simplifying and eliminating waste from the process. From Deming I learned that pushing numbers was a destructive practice. I learned that reducing cycle time is an essential element of process improvement. Today, cycle time is as important as it ever was, but how it is reduced is still of paramount importance.

If my way of managing my area was challenged, and it often was, my defense was that my "system", which really means my management style, achieved results. I achieved results in my unenlightened days, but the results that I achieved did not even compare to what I would have achieved if I had gone about managing the right way. If my department was able to beat the numbers it was my management that deserved the credit. If they failed to achieve acceptable numbers, it was my employees that failed. How wrong I was!

I was the consummate taskmaster back then. Close supervision of workers was the accepted management technique, and I was good at it. When they passed muster and made their numbers, I praised them. When they missed a step, I wanted to know **who** the culprit was, not what the problem was. The solution was always associated with somebody screwing up. After all, I hadn't missed a beat, they had. I had done everything I could do – the fault was with my employees.

By paying attention to what Deming was saying in his works, I finally learned that I had been wrong all along. My employees were not at fault, I was. I was not providing them with proper leadership; I had caused their failures. Shingo's works encouraged the reduction of setups and the simplification of my process. Goldratt wrote about how to deal with constrictions and I learned from him. Taiichi Ohno showed the world how to use Kanbans to control WIP. The pioneers in process excellence taught me that I was totally out of whack in my thinking. I paid attention and I learned **the real and only truth**. The truth is that it is the **combination** of all those things, or rather the **right** combination that makes the real difference. How these values and tools are applied makes a huge difference. FCP **is** that right combination and application!

Almost magically I was transformed into a different person. I came to believe in the concept of moving product through a process quickly and efficiently with as little waste as possible. It did not happen instantly, but in stages. The more I learned, the more I wanted to know. What a concept: being able to move product faster without trying to work faster. I found that it is possible to speed up without actually trying to speed up, get more output without focusing on output and improve quality without focusing on improving quality. The knowledge changed my life.

In my studies of Deming and Japanese management techniques, I learned that management and workers should be a team, not just people with different jobs. This concept was not a Japanese monopoly, but something that could and should be done here in the States. But that was not all of it. I came to realize that JIT principles worked with virtually every element of the factory, not just manufacturing. Manufacturing, Engineering, Training, Production Control, Sales, Marketing, Facilities and even Human Resources should think of each other as teammates with a common agenda and not as separate elements with different priorities. The values that are good for manufacturing should be embraced by all elements of the company.

Having a variety of beliefs and conflicting values and measurements in **manufacturing** is destructive to process improvement. With JIT, all elements of production should come together as partners through a belief in identical values. CFM taught this notion also. It is important that **all** elements of a factory unite against the competition, not against each other. The people on our side of the rope should all be pulling together. It does not work when one department makes itself look good at the expense of another department. A department is successful when it contributes to the success of all departments. That applies to production and support. Common values and common measurements lead to success.

One department can meet its goals and objectives to the detriment of all. It is not only possible, it happens all of the time when there are opposing priorities. Maintenance wants to do PM's (Preventative Maintenance) on machines during prime production time. Engineering wants machine time without prior notice. Manufacturing refuses to give time to Maintenance and Engineering even though they could afford to do so simply because they have not "scheduled" the time. One manufacturing department decides to bring people in on weekends without any coordination with others, even though they will be creating a WIP bubble and dumping too much work on their internal customer. These are some examples of departmental misconduct that is always destructive to the big effort and what is **really** important: beating the competition. Instead of making life miserable for our competitors, we make life miserable for each other.

I once wrote a white paper that spoke of the dangers of what I called "Philosophical Fragmentation". I know what you are thinking … Philosophical Fragmentation, for Pete's sake? Well, in my own defense, I really believe that fragmented or multiple opposing manufacturing philosophies do nothing but create havoc in production. You think push, I think pull. We are motivated by different objectives, and we reward our people based on their ability to react to external customer needs quickly. If I believe in **pushing** product to my internal customer to enhance my output report card, but my customer wants only what is needed when it is needed, I can make my customer look pretty bad by dumping product on them at the wrong time. They will have production problems and I'll come out of it smelling like a rose. However, if I provide my customer with predictable deliveries and great quality, we **both** benefit.

If I have the ability to produce more widgets per hour than my customers, I can bury them and continue to bury them and I will look like a hero. Meanwhile, they can fail in the eyes of my management. Why would I do this to them? **Because I can!!!** Also, I can hold back product and give my customer less than they need all day, starving them for work and then I can dump a bunch of work on them towards the end of the day. In tomorrows production meeting I'll look like a hero and they will have failed in the eyes of management. That **will not happen** in a FCP factory. In a FCP factory, I will be exposed as the problem, not my customers. **Better still, in a FCP factory, I would not *want* to do those things to my customer!**

Fast Cycle Production, despite its name, does not insinuate that it is necessary to hurry up in order to achieve low cycle time. It was so named because of my belief that cycle time reduction, and how it is achieved is essential to any business. The way to reduce your cycle time is to reduce the portion of your cycle time that is non-value add.

I emphasize this point later when I talk about **Manufacturing Cycle Efficiency (MCE)**, also known as **Process Cycle Efficiency (PCE)** in service organizations (Part 2, Chapter 2). If you think that achieving low cycle time means hurrying up, you are missing the point. You are not just missing the point; you are not even in the same neighborhood as the point. I teach that the only way to go is to eliminate cycle time through the elimination of non-essential process steps that slow the product down and adversely affect product quality.

Quite often, the greatest opportunity for cycle time improvement is to simply eliminate inconsistency in product transfer from one work center to another. This notion is different from the belief that all that has to be done is to find waste and eliminate it and thereby improving cycle time. Managers often fail to understand that increasing the frequency of needed deliveries from one work center to another has a number of positives associated with it, and inconsistent delivery of product has nothing but negatives associated with it. Inconsistent delivery of product is often overlooked when people are trying to eliminate waste. If people overlook the frequency of product delivery from work center to another, they are missing a lot of opportunity. TAKT Management helps to promote timely delivery of product. In many of my projects, achieving discipline with respect to product transfer provides the implementation team with its greatest challenge, and ultimately with its greatest benefit.

Now here's a news bulletin: Simplifying a process and achieving consistent product transfer has a profound effect on product **quality**! This is possible without maintaining a focus on quality. When the product moves predictably, with as few delays, distance and production speed bumps as possible, the result is productivity **and** an improvement in quality. Fast Cycle Production does not take quality for granted. It promotes a quality-oriented environment that considers quality first in the formation of solutions. No decision should be made that can possibly adversely affect quality. You do not have to compromise quality in order to meet heavy demand. It is possible to create solutions that never threaten quality. **Quality and productivity can be achieved without compromising one or the other.** When you have established a smooth, efficient process with little waste, improved quality is a derivative of the simplification effort, not something achieved through focused intent.

As I said, I came to believe in FCP values so much that to me they really and truly are values in the truest sense of the word. They are as ingrained in me as the values that guide my moral life. I have always encouraged my students to believe strongly in these values, not just accept them. I encourage them to find solutions to their production problems that fit within that value set. I have spent much of my career as a consultant working to convince others of the worth of these values. I always promise them that they will certainly be successful if they maintain these values. Remember, **every** client of mine has improved their process and its business dramatically. The only problem that my clients have experienced after a successful implementation was to continue to do what made them successful. As long as they maintained good production values, they continued to prosper. When they abandoned the values that made them successful, they lost the gains that they had achieved.

At the risk of being thought of as cocky and opinionated, which some say that I am, I am convinced beyond a doubt that my methodology can help any company out there to improve their productivity significantly by any critical measurement. But the real truth is that it really does matter **who** is in charge of implementing the project. Only true believers should be given the task of leading others in an implementation effort. The true believers had better be ready for some serious pushback from all directions, but they should not be deterred from moving forward. Sometimes, this resistance comes from surprising sources. I have seen it come from just about everywhere, including the people that ordered and paid for the help: senior management (Part 8, Chapters 1 and 2).

Management at all levels will sometimes look the true believer in the eye and say that they are all for the improvement plan and then fail to be supportive when their support is really needed. This may come as a surprise to many, but some managers are more concerned with their own political position (how they are seen by their own managers) than they are with achieving process excellence. Some of them will fold like an accordion when it comes time to defend a rightful position when the challenge is coming from a formidable source. On the other hand, others will hang onto a faulty position even when logic and data show that their position is wrong, which is an equally destructive practice. They cling to their position like lions holding onto raw meat. Fortunately, there are managers who listen, evaluate and maybe even challenge, but once they are convinced that what they are doing is the right thing to do, they become strong, unshakeable advocates. They are in the game for the long haul and work to arrive at solutions that are within their value set. This is the kind of manager that you want in your company and your department.

Then there is the pocket veto (lack of support through inactivity) that hits when you least expect it. People fail to support you by doing nothing when their support is needed. If they would have given you the needed and rightful support, you could have moved forward. As it is, you are stuck with a delay because of their inaction. Waffling managers might also look you in the eye and say that they are on your side, but they do not do a thing when operating rules are violated. They act as if they just do not want to get involved. I have seen managers of this ilk in just about every project. This type of manager lacks the necessary mettle to work on a productivity improvement team.

Occasionally, misguided shift managers will take actions that allow them to get good production numbers at the expense of other work centers. They know that they are breaking the rules, but they give orders to their people to bypass the rules "temporarily". An example of this is cherry picking (working only on the easiest product) in order to increase output numbers. This kind of egregious activity almost always results in problems because the other shifts have no recourse but to work on the toughest product, instead of maximizing their output by working a good mix. In cases like this, one department prospers but the company is hurt.

Another example of selective rule breaking is to create a parochial priority system in the running of orders, rather than working all orders in a first-in, first-out manner. Some departments do this to minimize the effects of setups, not realizing that because of this practice other orders may miss their due date. When setups are a problem, somebody should be working on ways of reducing setup times (See Setup Reduction – Part 5, Chapters 1 and 2). By the way, the examples of cherry picking that I've given do not just happen on off shifts; I've seen the day shift people practice cherry picking also. Usually, however, managers on the day shift know that they are scrutinized more than the off shifts and are therefore more inclined to avoid the practice of cherry picking.

After significant gains in productivity have been made through process simplification, the gains can be sustained and expanded for an indefinite period of time with good continuous improvement tools, but it is always wise to make sure that there are people in place that will stick to their values or there will be a serious erosion of productivity very, very quickly. That erosion does not need to happen, and will not, if you stick with the one that brought you to the dance.

I have learned from experience that if continuous improvement is your priority and you continue the theme indefinitely, you will improve your bottom line and your ability to gain market share indefinitely. If you do continue to improve, you can realistically expect to eventually become a true world-class producer. A little far-fetched, you say? Well, I know for a fact that becoming a world-class producer is not just a pipe dream; it can certainly be a reality for every producer of goods **and** services if they have the fortitude and conviction to work for it.

There Are Clients - And There Are Clients

The truth is, I have never received a call for help from a client that was consistently able to beat its competition in quality and delivery. They were either trying to capture more market share or they were in danger of losing what they had. When I say that they needed help, I do not just mean that they **thought** they needed help; I mean that they openly **admitted** to needing help. The difference between admitting that help is needed and failing to admit to the need is significant. There are a variety of ways that a production company can need help. I am talking about needing help to make a quality product and get it out the door faster while at the same time using fewer resources. The true definition of productivity is to produce more with less, and to produce more quickly without cutting corners.

Some clients could sell more if they could only produce more. The orders are there but the ability for their processes to produce the product is not. They cannot afford to spend more money on their processes or hire more people because such spending would adversely affect their profit margin. In order for these producers to survive and prosper, they must get their product to their customers quickly and be able to take on new customers without sacrificing quality or delivery.

I have been asked by a few companies to come in, look things over and tell them what needed to be changed. In each case, I assumed that they already knew that their processes were flawed when they called me in; they just didn't know what to do next, or that they were just trying to determine whether or not I could help them. Truthfully, I have never been very shy about finding and talking about opportunity. Usually, it is only necessary for me to tour the plant or business just once and I can spot the opportunity that is there. I have been told that it is a gift. I believe that it is because I look at the process and measure it against my production value set. If I see something that violates my value set, I have probably seen it before and know what it takes to fix the problem. If it is a gift, it is a gift that was developed over time. Most important is the fact that "it" is a gift that can also be transferred to others. Others can do what I do if they look through a trained set of eyes with the proper value set. Basing decisions on sound values can help anyone to see both opportunity and solutions.

As I said, the type of company that usually asks for my help is one that has a good or great product and could sell more of it if they could only produce more, and produce it faster. Occasionally, it is a company that, in order to gain a larger market share, has over committed and accepted orders that it knows it will have trouble meeting. I have worked with companies that have lost market share because of serious process problems and had let some of their people go because of their problems. It is difficult to improve the process of a company that is looking to reduce its workforce. If the company is downsizing, it is not usually concerned with its processes. This is unfortunate because in every case of downsizing, I think that process excellence should be high on a list of priorities for that company. Finding ways to do more with less should be their battle cry. Usually they are just trying to do less with less.

Since downsizing usually happens very suddenly there is not normally an emphasis given to process improvement just prior to the layoff. Once the decision is made to reduce a certain percentage or a given number of employees, the focus is completely centered on just making the layoffs happen smoothly. If the processes were in trouble before the layoff they will be in even worse shape afterward when many knowledgeable people are gone. Obviously, if those same companies would have paid more attention to their processes in the beginning, when things were going well, they might not have found themselves in such a precarious position afterward. The point that I am making is that process improvement is important regardless of the growth capabilities of the company. When the orders eventually come in and the need for more employees becomes a reality, the process is ready.

Some clients keep their cards close to the vest and fail to communicate their needs to the consultant. I prefer clients who are open and honest about their needs and are looking for a partnership. This relationship saves so much time and cost in the long run. I have been fairly fortunate with my clients, possibly because I am fairly blunt and know which questions to ask and possibly because my clients really needed help and knew it. I've seen all types and certainly prefer the honest, open types. When I think that a client is being coy or evasive, I have been known to press the issue until I determine whether the client and I can form a partnership. If we cannot, I kind of think that maybe, just maybe, we were not meant for each other. As a result of this rather admittedly inflexible approach to selling myself, I have lost a few contracts. Such is life!

I've listed below some of the jobs that have worked very well for me. In **each** case, client expectations have been greatly exceeded: The client got more than they bargained for and were extremely happy with the results. In other words, the project was a huge success:

1. Company A had a great new product that was brand new and it had to grow its process to meet the anticipated demand. Its concerns:
   a. How many shifts do we need?
   b. How many new machines do we need?
   c. How many people do we need, and where do we need them?
   Results: Through process improvement and minimal layout changes, only one shift and 2 fewer people on the line were needed. Output was improved enough to meet future demands without capital cost or new employees.

2. Company B had picked up some new contracts but was unable to meet the demand. According to its IE's, it should have

capacity but cannot seem to keep up with the new demands. Its concerns:
- a. What do we need to do to meet order requirements?
- b. What do we need to do to meet acceptable lead times on orders?
- c. Do we really have the capacity we think we have?

Results: Through process improvement, output and cycle time met and exceeded requirements without capital equipment cost or new employees within 30 days of implementation.

3. Company C had a new product line but its product lead-time was too long. If it could cut its lead-time by one third, it could pick up a 15% market share. Its concerns:
- a. Can we reduce cycle time and still maintain product quality?
- b. Can we reduce cycle time without buying new equipment and hiring more people?
- c. Can we reduce cycle time AND improve output?

Results: Process analysis and simulation were used. Lead-time was cut in half, no machines to buy and output AND cycle time improved by 50% within a month. Quality of product improved even with higher output. Cycle time eventually reduced from 7 days to 1.

4. Company D had a mature product but the demand for the product had dropped significantly although the level to which it had dropped would remain stable for some time. Although the output requirements were reduced, the price of the product presently being produced was higher. Its concerns:
- a. To what levels can we reduce our work force and still maintain good productivity?
- b. How much can we shrink each department without losing more capability than we can afford?
- c. Can we downsize and still maintain the capability to handle an unforeseen increase in demand? How much can we downsize?

Results: Instead of hitting all departments with an across the board reduction, simulation modeling revealed that it

was it was more feasible to reduce 3rd shift about 75% and keep 1st and 2nd shift intact. Those remaining on 3rd shift were highly skilled workers that kept critical machines operating and performed preventative maintenance. Cycle time was reduced on 1st and 2nd shift through process analysis and simplification. The net result was that output loss was minimized with an improved cycle time. In addition, this company was poised to take advantage of an upturn in business when it arrived.

5. Company E has been told by Sales that its on-time delivery and customer satisfaction levels are so low that, if it did not improve, they will probably go out of business. Its concerns:

    a. It had hired so many people to meet requirements that it was losing money on the product that it did ship. How can it reduce its cost per unit?

    b. Even though it needs serious overtime to meet the demand, its machine utilization is low. How can it improve its utilization?

    c. In order to meet demand, it has to maintain a large finished goods operation. How can it build to order (on-time) instead of building to stock?

Results: This was a tough one that required a full FCP implementation! TAKT Management and Kanban control, along with some serious elimination of waste through process mapping and simulation modeling completely turned the situation around for this plant. Product started moving steadily throughout the plant and overtime was almost completely eliminated. Cycle time was reduced from 27 days to 8. On time delivery was no longer a problem with a 100% on time delivery rate happening within 60 days of implementation and sustained for a year. It was able to pick up new customers and solidify its previous customer base because of a tremendous customer satisfaction rate. Although it was on the brink of destruction, this plant made a dramatic turnaround. This plant went from worst to best of breed within 9 months!

The results have always been great, and the improvements did not just stop after the project was finished. In every case, the client has been able to continue to improve long after I had finished my work with them. My goal as a consultant has always been to make my clients self-sufficient. They have been able to achieve self-sufficiency because the seeds of continuous improvement have been planted. All they had to do was maintain the values that made them successful. Maintaining the proper value set is easier said than done.

Even though the gains were huge, I found that there was some erosion of values eventually **in each and every case**. I was called back by clients on many occasions and found that recidivistic behavior had all but eliminated the gains that were realized during and after the projects. A few of my clients had some serious sliding as a result of values erosion; in fact, I know of several who dropped everything that made them successful and went back to a condition very close to what they had before the improvement project. I have spent years trying to figure out why people do not continue to do what has made them successful. If you were doing well, why would you return to ways that you historically know are counter productive?

Fixing something is one thing, but why would you want to undo everything that you have worked so hard to achieve? I do not yet have the answers to all of these questions, and although I now have a system that has built-in protection against something like that happening, there is no system that provides perfect protection because bad or complacent management can trump any accomplishment. Looking back on it all, I now see that in virtually every case where the regression occurred, the cause was managerial indifference. The determination that pushed them to succeed was replaced by complacency. In a few cases, the top manager(s) were replaced by people who did not understand or appreciate the new system or the tools being used.

Fast Cycle Production developed into a system that **always** works. It always works because its foundation is based on good, solid values. Values Based Management (VBM) always works when the values are maintained. For many years, I would look back on clients and see them fall apart even after they had made such dramatic gains. Once I found the elusive, secret ingredients that could enable a system to be easily maintained, I felt I had to share them with others. I was not just interested in getting work; I wanted the system to have longevity.

When you have a continuous improvement system that just keeps on getting better and better, you do not want to see the good times come to an end. You want to be prosperous forever! I do not just believe Fast Cycle Production to be a good productivity enhancing system; I consider it to be the best of them all. It is the best of them all because it is the best application methodology of time tested values. It simply cannot fail **unless it is abandoned**.

Chapter 4 - Just In Time Manufacturing – Will it only work in Japan?

Back in the early 1980s, when the Japanese were getting a lot of attention from American manufacturers because of the success of Just In Time (JIT) as a manufacturing philosophy, Toyota was making better cars than Detroit and its sales were skyrocketing as a result. The Toyota was not just better in price; it had better quality than its American competitors. But it was the "Japanese" that were making these cars! The Japanese were known as copiers, not innovators at that time. Besides, Americans believed that the Japanese workers were different than American workers. We believed that they would work for less and work harder than Americans. **"This was part of their culture! They had no mind of their own! They were submissive by nature!"** We tried to explain their success in many ways. Boy, were we wrong!

Back then, there was a lot of talk about how this new phenomenon called Just In Time Manufacturing was just a cultural thing that "would never work with American workers." American workers required "pushing" to achieve good results. American managers, especially in the manufacturing industry, seemed to believe that only "good management" produced good results from the worker. I have vocally maintained that the American worker is as good a worker as you can find anywhere on earth. I was right then and my statement is still true to this day. But good management *is* needed in production. It was true then and it is true now. But there is a hitch to that statement.

Those who thought that management is the key to success were right in one respect –good management DOES produce good results; however, I think that they were wrong in their definition of what good management really is. Japanese managers clearly were getting good results from their people, but the good results were not a result of close supervision or strong-handed management styles. The Japanese believed that workers should be involved in the production process, and that the workers were part of the solution, not the problem. They involved the workers in meaningful ways that produced results. The Japanese managers worked to earn the workers' commitment **and** their loyalty.

That old cliché about commitment and involvement rings true. It says that with ham and eggs, the chicken is involved, but the pig is committed. In the U.S. we thought of Japanese managers as authoritarian. As a group, they were clearly not of that ilk, at least not those managers who practiced JIT. They involved their people in ways that led to productive improvements. You see, they did not just **allow** JIT to happen; they **made** it happen. Think about that last statement for a while and think about what management commitment really means in a production environment.

The JIT managers believed that workers did not get up in the morning with the express purpose of screwing up. They believed that everyone has a strong desire to succeed at what they did. They did not think of their workers as lazy or as having bad attitudes. Their value set led them to believe that all workers truly wanted to be successful in whatever job they held and that coercion or threats were not the way to make them successful. They were simply people with different jobs not people to be looked down on. Could it then be said that the way the Japanese managed their employees was the result of cultural differences? Could JIT work in our culture? When I said that we have the best workers in the world, how could those workers fail if we gave them the best system in the world? Answer: They can't, if we also gave them the best managers!

American managers certainly have the capability of being outstanding production managers. They have everything going for them that make them capable of even surpassing everything the Japanese did. They have good workers and good brains. What then are they missing? In my opinion, they are only lacking in the areas of **sustained commitment to solid production values**. When the battle is joined, they are right there in the thick of things, doing what is right. Afterwards, when the main battle is over and the gains are much smaller, albeit just as significant, these same people who went to war to be the best, find something else to be excited about. They want the big gains, the big rewards and plenty to brag about. They don't seem to be satisfied with sustained growth at a lighter level. They forget about the process until things start to fall apart again.

I did not know all of this when I first started studying JIT, but I knew that JIT worked, and would work in a country other than Japan. Instinctively, I knew that it was not a cultural thing at all. I became an insatiable and almost fanatical student of JIT and all of its beliefs. I mostly studied the works of Shingeo Shingo, who I considered (and still do) the most brilliant industrial engineer ever. I read everything he wrote. I read every book that I could find on the subject looking for the common denominator that made JIT so successful as a manufacturing philosophy. I went to every seminar that my company would let me go to and some that I financed on my own. My best source of information on the subject turned out to be the public library. At the time, I was on a personal crusade to learn, not to develop a philosophy of my own. The latter came much later.

One thing that I discovered was that it is never the nationality of the worker that matters; it is the leadership that the workers get, or do not get. It does not matter whether managers are American, Japanese or whatever. What matters is **how** they manage their people. If their values are good and consistent and they provide **sustained** leadership to their reports, they will do well regardless of their nationality. JIT could just as easily been created in the U.S. if the need to succeed had been as intense in America as it was in Japan after the war. The Japanese had no choice but to succeed. They could not hope to succeed against American industry unless they were the very best. They couldn't follow our way, so they created a way of their own. It worked!

My obsessive study of JIT was the beginning of my belief that manufacturing was indeed a science. However, I would later revise the term, **"The Science of Manufacturing"** to be, **"The Art and Science of Manufacturing"**. The creativity required to deal with unforeseen occurrences on the production floor is the reason for the addition of "art" into the term. In manufacturing, you have to react quickly and correctly or you can have some serious problems. It takes creativity to come up with solutions that fit a particular situation and are consistent with accepted values. It is so easy to come up with a band-aid solution that will probably fall apart before long. The perfect quick decision is a work of art.

JIT was not made just for manufacturing. It has been used effectively in just about every industry. The beauty of this philosophy is that it can be used wherever there is a "product". I have used JIT effectively in virtually every industry, including the service industry where I was amazed to discover that JIT values are not only usable, but also effective. I have used JIT values when working with banks, lumber companies, chemical companies and even homebuilders.

At one time I thought that JIT itself was the philosophy that was the true answer to the science of manufacturing. JIT started out in an automobile factory and eventually spread to other industries. It was able to expand because the values were not just applicable to automobile manufacturing. The values were so solid that they were ideal for any type of business. It would be many years before others and I would come to realize that JIT was not just for manufacturing, in fact, some still believe that JIT is tied at the hip with manufacturing.

Initially, when JIT was brought to America, it did not have the same exuberant acceptance that it enjoyed in Japan. JIT was viewed as a Japanese thing. It was believed that American workers just were not suited to this type of system. It was insinuated and even bluntly stated that American workers were inferior to Japanese workers because American workers didn't work as hard as the Japanese worker. American workers only wanted more money or better benefits while the Japanese worker was dedicated to their employer. It is true that in those early days, Japanese workers considered their work as an essential part of their lives, and it is also true that American workers wanted more pay and better benefits. But just because you want to improve yourself does not mean that you are not an outstanding worker. American producers were wrong when they attributed the success of JIT to culture. By the time that they woke up, they had lost some very serious market share. Toyota was not yet in the same league as GM or Ford at that time, but they had established a very strong foothold. As we all now know, Toyota is doing very well, thank you. It all started with JIT over 30 years ago.

If JIT values were sound and were applicable anywhere, why was there so much rejection of JIT by American corporate leaders in the 80s? Also, when it was accepted and successfully implemented by some American companies, why were they unable or unwilling to maintain the gains, even after achieving very substantial improvement to the bottom line? Was it the difference in American and Japanese workers? If not the workers, then what was it? The answer to this question surprised many people.

The difference between the sustained Japanese successes and the American inability to sustain success was totally attributed to management differences. Japanese managers knew how to manage in a JIT environment and American managers did not. American managers did not readily accept the JIT values; **they only demanded that their people adopt the Japanese values**. They had (and still have) the perception that their only responsibility is to manage, not to be an advocate of the JIT values. They were and are very hypocritical, values wise, a sad but true fact. Their rule was to, "Do what I say, not what I do."

Don't take what I've said as a denunciation of American managers, because you would not be accurate. Our country has the most talented and the brightest managers in the world. In terms of ability, I would stack American managers up against the best in the world. The problem is that American managers prefer quick dramatic gains that enhance their political position within their organizations. JIT and CFM gave them these dramatic gains only during the implementation period and the gains lessened as the company moved into a continuous improvement mode. When significant improvement starts to go away, many managers lose their focus on continuous improvement. They seem to be hung up on dramatic results, not smaller but consistent gains.

As to the difference between Japanese and American workers then and now, I know from experience that American workers can be as good or better than any other worker in the world when they are managed by capable leaders. There are always going to be labor issues, and there are always going to be a small minority of people who are lazy and nonproductive, but this situation represents a very small percentage of the working population and is true with workers throughout the world. American workers are no different than their foreign brethren in one important respect: They truly want to be successful and want to be part of a successful effort. The definition of success varies greatly among many people. Success to one person might mean retiring early. To another person, it just might mean doing a good job and being part of a successful company. Still, to another person it might just mean being recognized by his or her boss and peers as being good at his or her job. Workers enjoy being successful – It makes them feel good about themselves and improves their self worth.

Our Business is Unique – Baloney!

Prior to every application of FCP, the same question always comes up: "Have you ever worked with a company that does what we do?" Another statement that is almost always made is "We are a very unique kind of business. Many rules that apply to other types of businesses probably are not valid with us." Clients always insist that their business is "unique". The fact is that they are all alike in one very important respect: There is opportunity for improving every discrete event process. Cycle time improvement is important for each and every process out there. But it makes a great deal of difference **how** you reduce cycle time. Never cut corners – eliminate obstacles. You will learn a lot more about cycle time reduction later on when **Little's Law** is covered (Part 2, Chapters 1 & 2).

The main point here, as you will hopefully discover, is that a methodology that works for manufacturing can also work for other industries, including the service industry. It will also work for support areas of manufacturing companies. Of course, not all of the tools can be used in all industries, or even all manufacturing plants, but the values can and are used in virtually all industries. Banks, a lumber company, distribution centers, producers of pigments, circuit board (panel) manufacturers, electronics manufacturers, medical equipment manufacturers and a company that manufactures homes have all benefited from the values contained within Fast Cycle Production.

When I said that FCP was used with a company that manufactures homes, I did not mean that they produced the so-called "mobile" home. The homes that I am talking about have foundations and are made of traditional building materials. The business actually consists of handling everything for the eventual homeowner such as:

1. Buying the land
2. Building the home to the customer's requirements from an assortment of plans.
3. Providing all of the interior amenities to the customer's requirements including carpets, drapes and cabinets.
4. Financing and processing the loan.

In short, they are a manufacturing company, even though their "product" is homes. If their processes are well managed, they will make more money. If not, they will lose money. In this sense therefore they are no different from any other manufacturing company. All producers of products can benefit from good production values contained within their processes.

Remember, no matter how popular your product is, out there somewhere is someone who can find a way to produce it quicker, at less cost and possibly with better quality. No business is safe from this fact. No matter how much money you are making, if you improve your process, you can make more. Good products do not remain popular forever and making as much profit as you can and holding down your process costs is at least as necessary during the growth period of your product's life as it is later. What you lose from inefficiency is lost forever. Opportunity missed in production is opportunity lost. Profits lost are also lost forever. You may make more money later, but what you lost is gone. If the window of opportunity is missed, another company may gain a significant amount of your market share.

Greed Can Be Costly

Working for a successful company can sometimes lead workers to presumptuous conclusions. They can come to believe that since their company is successful and is making a lot of money, they ought to be paid more. This becomes a problem when what the workers think is fair exceeds the worth of the job. Labor unions are in the business of getting raises for the rank and file. The more raises and benefit increases that they can negotiate, the more that they justify their existence. Unfortunately for the companies, excessive pay and benefits can eventually force them out of business.

Every job, no matter how important has a maximum value. Some workers believe that seniority is the only prerequisite to a pay raise. Once that maximum value is reached, the worker is left with a dilemma: In order to progress further in responsibility and pay, there must be preparation and qualification. If a person expects more responsibility and pay, they can always prepare themselves to be qualified for the position that they aspire to. Workers cannot expect to inherit a job that they are not qualified for based on their seniority just because that job will provide them with more money. Seniority may be considered when several workers are qualified for a position, but seniority is not what usually gets you the job.

Workers at any level, from assembler to plant manager to CEO have options when they conclude that they are not being fairly compensated. They can prepare themselves for the next level through education or experience, move to another company or bid on another job for which they are qualified in their own company. Sometimes the only way that a person can get promoted is to move to another company. But whatever their course of action, they cannot expect to really be successful in their next level unless they are successful at their present level. Sitting back on their haunches and waiting for something good to happen to them is not a course of action I would recommend.

Looking at the other side of the coin, I know from experience that sometimes companies try to get something for nothing from their people. It is the nature of the beast called business I suppose. There are hiring policies out there that would surprise even the most pro-business people. Some companies absolutely refuse to pay people fairly even though it is within their ability to do so. This destructive policy usually ends up forcing them to suffer from high attrition rates. People just stay with them until they find something better. Some companies serve as training grounds for other companies that compensate fairly. Employees will gain experience at low pay and move on to another company that pays well when there is an opening.

Companies with high turnover rates don't really save money with their low-pay policy. They probably never will know what they have lost because of their policies. In some of the companies that I worked with, adequate pay was a big issue with some departments. While a team of people was trying to make something good happen with the process, it had to deal with people who were under compensated and were just waiting for an excuse to leave. They could care less about improving the process; Salary was their issue.

It seems to me that industry is always in one of the following states with respect to workers: There are so many available workers to choose from that they can get away with paying less, or there are so few that the company has to "hire the warm body" and hope for the best. The last decade has had more of the latter than the former, although the climate is changing at this writing. Companies are starting to recognize the importance of company loyalty and how to achieve it with their regular employees.

The use of temporary employees is growing because companies are trying to save money on benefits. One manufacturing company that I worked with had a 40% temporary worker population throughout the plant. They also had a six to nine month maximum for all temporary workers in some departments; however, because the training curves in certain departments were about as long as the workers tenure, the workers were leaving at about the same time they were reaching their optimum skill level. Still, this company felt compelled to staff these highly technical areas requiring difficult training with temporary workers rather than assign them to less technical areas where the training time would be much less. I advised management in one plant to reduce the number of temps in the technically difficult areas, but little attention was given to that advice for quite some time. Apparently, it was a corporate decision that was "managed" at the plant level.

During the early days of JIT, the Japanese did not have a worker retention problem. It was common for a worker to spend his or her full adult working life with one company. The company goal was to create loyalty. Employees were loyal to their company and did their best day after day. Companies showed loyalty to their employees by making sure that they were adequately compensated and enjoyed great benefits. Some of the benefits enjoyed by Japanese workers would make most Americans envious. We thought that they were getting less pay and benefits than we were ... wrong again!

On a family vacation to Hawaii some years back, I tried in vain to get a tee time at a nearby golf course because there were so many Japanese playing there on a company paid vacation. I talked with several of them and discovered that almost one hundred of them had been given a paid vacation there because of various contributions they had made to their company. Examples of these contributions included perfect attendance records and suggestions that resulted in improvements. How many companies can you think of that do that, especially in today's economy? Those old days are probably gone perhaps for a long time, even for the Japanese, but at the time it really worked. I wouldn't be surprised at all to see that kind of thinking return some day.

As you read this and think about your own company, don't you think that loyalty should be a two way street? I know of some American companies that had similar beliefs during the 80s. IBM and

ROLM are two with which I have personal experience. IBM in the late 80s proudly declared that the "Dignity of the Individual", was important to them. Their actions proved that they meant what they said. ROLM simply and proudly said that ROLM was a "Great Place To Work", and from personal experience I can tell you that it was true. I am sure there were more companies that fit the same mode. The Japanese were not alone in their family philosophy with their employees. Throughout my career, I worked with companies that asked for loyalty, but then had a big layoff right around Christmas time. Christmas is right at the end of Q4 is it not?

Chapter 5 - It Takes A Village ... I Mean, It Takes A Company

Although there is no <u>single</u> person who I can thank for helping me form the philosophies of Fast Cycle Production, there is one <u>single</u> **company** that had a great influence on the building of FCP principles and tools. That company is IBM. Back in the 80s and early 90s, IBM had the right kind of people and the willingness to seek world-class dominance in manufacturing. At IBM there were those who were hugely impressed with the Toyota success story and who wanted to transfer it into an American manufacturing philosophy and system that they could implement in their own factories. They were very serious about making this transition a reality. A group of IBMers formed a very select group of motivated and dedicated people that put together an outstanding manufacturing productivity-improving package called **Continuous Flow Manufacturing (CFM)**.

The CFM group started small and then blossomed into a sizeable group of individuals all dedicated to achieving process excellence. Their belief in their new product was sincere and passionate. They put together a training package and prepared themselves to teach CFM to other IBMers. Although there was disagreement among them over certain points of the philosophy, they debated and came to consensus on what should and should not be included in the package and how it should be presented. This group of IBMers was not a rag tag group - these were knowledgeable people with outstanding backgrounds that could articulate their philosophies. Most important to me was that I saw in this group total dedication to their cause and belief in the CFM values. They might debate issues, but they were united when it came to the values contained within CFM.

When I wondered out loud one day why they decided **not** to call their package JIT, since the basic values of JIT were nearly identical to CFM principles, I was quickly told that CFM has more in it than just JIT, in fact they considered it better than JIT because it could be implemented in a factory much more quickly. They had done their homework in their search for a better manufacturing system. When IBM came up with CFM, they were not just copying JIT from Toyota; they were taking what the Japanese started and then working to make it an "Americanized" version of JIT. To create CFM, they kept many of the tools and values of JIT, and added some of their own. Most important to me was that they had developed a very clear and repeatable implementation philosophy. They didn't just copy JIT; in their opinion they improved upon it. I have followed that same way of thinking with the creation of FCP.

IBM invested heavily in the development of CFM. They followed up their investment in true IBM fashion by working to implement CFM into every IBM manufacturing plant - everywhere. When I first joined the IBM Consulting Group, which was about a decade after its conception, I worked with the CFM team to sell CFM, not only to IBM manufacturing, but also to other companies. CFM had been implemented in many IBM plants and the results were **all** successful. Let me reiterate this point: A CFM implementation **always** achieved successful results! Does this last statement sound familiar? Remember that I said successful results, not long-term results.

Through CFM I had found an American version of JIT and I loved it! I had hit the mother lode of manufacturing knowledge and I could see first hand how dramatic the results could be. I took part in several CFM implementations and the results were great each time. One of the implementation tools that we used in CFM was called Takt, derived from a German word meaning "beat". With FCP I have modified the CFM version of Takt and created **TAKT Management.** I credit CFM with the discovery of the best continuous improvement tool I've ever experienced: TAKT Management. TAKT and TAKT Management are covered in detail later (Part 3, Chapters 1 & 2).

Now, let's get back to CFM. As I said, it all started at IBM with a group of people who studied JIT as I did but who went further and worked to create a repeatable implementation package for their own plants. They created the complete package and continued to fine-tune it for years. These IBM pioneers put together a darned good package. Although I later modified much of what the designers of CFM came up with, I did not change the values they adopted from the Toyota model. Those were and are still valid. On a sad note, when the CFM group was expanded, several CFM consulting teams were created. Eventually these teams became competitive towards each other resulting in political battles between teams that resulted in the eventual demise of the CFM offering. The teams were disbanded not because of the inability to sell CFM to clients, but because of political infighting.

The creators of CFM used many of the JIT tools, like *Kanbans* and *Poka Yoke* and added some new wrinkles, along with a very well disciplined and documented implementation and training plan. They trained internal consultants at IBM who had one job: Implement CFM in every IBM manufacturing plant. The goal was ambitious, but it was certainly achievable, and they did their very best to achieve it. As you might guess, at that time there were a substantial number of IBM manufacturing facilities throughout the world. To have a goal of implementing CFM in all of them went far beyond being just an ambitious goal.

All of this work began before I even knew of the existence of CFM. I didn't know it at the time, but I would someday be one of the consultants that would be implementing CFM in IBM and non-IBM plants. When I did become involved with CFM, I considered it a perfect fit for me. I already believed in the CFM tools. At the time, I was more impressed with the potential of Takt than many of my peers. My opinion then and now is that Takt had more to offer than just being an implementation tool. I also believed that CFM Kanban theory was much too complicated, a key reason for its inability to stick once implemented. Overall, however, the system was an outstanding one. The CFM implementation methodology is the model for Fast Cycle Production today.

Although CFM had the same ingredients as JIT, everything was contained in an implementation package that took 6 to 9 months to complete instead of 5 years. In the case of Toyota, JIT took years to fully implement. FLASH: You can change from being a so-so or struggling company into a dynamic company with a world class manufacturing process in 6 to 9 months! The CFM team did a great job of documenting the philosophy to make sure that each implementation followed the same set of plans. All of the JIT values were in the package and were explained in a way that American workers could relate to them.

These IBM pioneers really had a good product and they knew how to sell it. CFM showed the possibility of really improving the production capability of any IBM plant. But here's the kicker: They only had to **implement** CFM; they did not have to stick around and make sure that "it" stuck around. Once CFM was implemented, the employees that were pulled out of manufacturing to help implement CFM in their plant went back to the jobs that they held prior to their selection to the "Implementing Core Teams". There was nobody left in place to make sure that CFM remained in place. This turned out to be a very deadly mistake.

CFM was true to its promise of bringing success wherever it was used. After each implementation, big gains were being realized in product cycle time, output capability and product quality. There were teams of internal consultants dedicated to implementing CFM everywhere. Although the teams were never able to accomplish their goal of putting CFM in every IBM plant, wherever they did so, they had great results. Each and every full implementation resulted in the dramatic improvement of every important measurement critical to manufacturing success. Most of the IBM plants on America's east coast went through at least one implementation of CFM over the years and there were at least several that had up to three implementations spread over five years or more.

But success can be a fleeting thing. The honest truth is that in **every** single case of a successful implementation, the CFM discipline only stayed in place for about six months to a year. After that, the plants went back to doing whatever they were doing before CFM. When I say that they went back to doing what they were doing before, I mean that they went back to losing productivity. Whenever CFM went away, management went back to being numbers oriented. Along with being numbers oriented, the plants returned to the practice of prioritizing orders. Let me summarize: When they were successful, they were process oriented; when they were not successful, they were numbers oriented! Is there a light coming on over your head right now? There should be.

How can something like this happen? How is it possible that a plant could improve so dramatically and then allow their success to just flitter away due to neglect? The workers on the line absolutely loved CFM! Their jobs were simplified, there were fewer meaningless tasks and they were getting more product out the door. The rules forming the discipline also made sense to them because they were based on what was right. CFM was easy to understand and offered them consistency in their jobs. Many of them called it "Common Sense Manufacturing".

Here is another tidbit that should get you thinking: During each implementation and for a period of time afterward, the wasteful process steps were removed because they made no sense. That is why the steps are called "waste". Soon afterward, these same wasteful steps returned through managerial and supervisory promulgation. A memo to count something twice would be issued by a supervisor or lead person. Workers would collect data for the engineers rather than have the engineers take the data themselves or arrange for those not engaged to take the data. Maintenance would interrupt work during prime production time rather than scheduling maintenance activity during off hours. Over time, the accumulation of these activities took their toll and CFM disappeared.

Think about what they gave up: they were once able to make more product, their jobs were simplified and their quality always improved. "Hey", I thought, "If it is so great, and everybody loves it so much, how could they let it slip away?" What could have been done to keep it in place? The Americanization of JIT was certainly an enigma. Just knowing that it disappeared was not good enough for me. I just had to find out why it never stuck. I knew some of the reasons, but was there just one thing that caused it or was it the aggregate? I thought that if I could just find the reason that it always disappeared, maybe I could find out how to make it stick. I searched to find the adhesive that could keep it together. The answer was quite simple, as I would one day discover, and it could be summed up in a two-word phrase: Sustained leadership. To expand the phrase slightly: A lack of sustained leadership. But how do you keep sustained leadership in place? Same problem.

I did eventually discover the adhesive that could keep a system like CFM or FCP in place, and I will share it with you, but first I want to go back in time before I first discovered IBM. Obviously, I knew that they (IBM) existed. I just did not know that they were going to buy the company for which I worked. Furthermore, I did not know that they would buy the company for which I worked for and then sell us. Who do I mean when I say **"us"**? Well, I am talking about ROLM Telecommunications, the very best company that I have ever worked for – that I didn't own, that is.

ROLM – A Great Place To Work (For a while)

ROLM was a great place to work – they even used that as their advertising slogan: "ROLM – A Great Place To Work". People there worked hard and got a lot done. It was a dynamic company that, in many ways, was way ahead of its time. As a manager in ROLM, I actually budgeted for things like picnics and beer busts for my employees. If you had a twitch when I said beer busts, can you imagine what IBM thought of it at the time? In defense of our practice, I can say that we worked hard and we got a lot done. We were what you would call super successful. Our stock just kept getting better and better. We even had free coffee in our break rooms. Our goal was to hire the best and the brightest. I like to think that we did just that.

We had a product that was at least 5 years ahead of the competition – a digital switch that could process voice (analog) and digital data simultaneously. AT&T did not have anything like it at the time, nor did anyone else. AT&T made archaic analog switches (yuk). It was like comparing a high-speed computer with a slide rule. This was the **first** digital switch ever built and there was nothing to compare with it. The competition had no answer to our great product. Nothing, that is, except a monopoly. If we did not have to deal with the power that AT&T could wield before they broke up, I truly believe that we could have dominated the selling of switches. Our product was that good, but our ability to manufacture that product was even better.

We had everything at ROLM: A great product, great people, and a great working environment. Many of us considered ourselves to be invincible in business. We had orders coming in like crazy. We had just built a beautiful manufacturing facility in Colorado Springs, Colorado and had our main manufacturing plant in Santa Clara, California right in the middle of Silicon Valley. We were invincible - we were dedicated men and women!

Speaking about men and women, I want to point out that ROLM was a place where workers **wanted** to spend their whole careers. We all loved working there during the glory days. We had good pay, security, stock options and a very good tax deferred savings plan. People were aware of the fact that they were important to the company. If I were to take inventory of our culture, I would say that it was a very healthy one because each employee felt that his or her effort mattered.

Hold it! Not everything was milk and honey in Camelot. There had to be something wrong. What we did not have was a product that was designed for manufacturability. In other words, our product was tough to make. Although the technology of the finished switch was top notch, the design of the circuit boards that went into the switches required engineering rework or design changes so often that the boards themselves looked like someone had dropped spaghetti on them. We had to have extra rework people to complete the engineering changes to the circuit boards because our regular rework people were already busy repairing boards that failed functional tests. The reason for the requirement for rework was because we ordered large quantities of printed circuit boards. Instead of having the producer of those PCB's make the design changes, we were forced to do the engineering changes (EC's) ourselves. It was then that we discovered the folly of ordering large quantities of raw materials rather than having a Just In Time delivery of material.

Bottom line, we had to really work to get quality boards loaded into our systems. As a result of all this, there were many of us who learned a great deal about the science of manufacturing, not to mention the importance of a good circuit board vendor and low WIP levels. To this day, I can honestly say that what some call my amazing ability to just look at a process and figure out what is wrong with it quickly and accurately is a result of my experiences at ROLM. That place was the advanced training ground for process consulting. By the time I left ROLM, I had seen it all and then some.

As a team, the ROLM people were innovative and dynamic, but we also did a lot of stupid things. Like other companies with lesser products, we made our share of mistakes. However, we always landed on our feet because we were so dedicated to success and to each other. We didn't know what it was like to fail. Because we had this great product that was selling like gold at $35 an ounce, we worked our way through every problem regardless of how difficult it was and, as a result, we were always successful. We just did not believe that we could be anything but successful. But ROLM was also a great teacher of what NOT to do in manufacturing. I was right there with the other misguided fools, doing things I should not do and actually getting away with it. We just did not know any better. As I said, we had such a great product; what did we need a great process for? You can't have everything can you? Yes you can! I just didn't know that yet.

In the middle of all of our success, IBM purchased us. The company that created CFM was about to bring its philosophy and its people to ROLM. What I found out later was that not all IBMers were believers in CFM. In fact, I found it very hard in those days to find an IBM employee that was knowledgeable about CFM. If this statement makes you pause with confusion, you should. How could a company that was attempting to revolutionize manufacturing with a philosophy joined at the hip to JIT possibly be hypocritical in its approach to manufacturing? You say, NO WAY? I say, WAY! CFM seemed to have its power base only back east. West of the Mississippi, it was hard to find except in Texas.

Shortly after they purchased us, some IBM managers out of the manufacturing plant in Boulder, Colorado, came down to Colorado Springs to join our staff. There had been a lot of restructuring at ROLM over the first few months after the purchase, but most, if not all of us, managed to continue on in our old roles or were given new jobs at our assigned levels. ROLM was growing and IBM saw our technology and theirs as a perfect marriage. Their hearts were certainly in the right place, but it would have been much better for us if they had sent us IBM managers who believed in CFM. Instead, the managers who came to us were old time manufacturing types. I think that Taylor was their hero. They certainly liked having a lot of WIP around.

Also, at that time, we were just trying to keep up with the large demand for our product. The philosophy was to sell, sell, sell and then let manufacturing keep the delivery promises. If there was anything that IBM marketing was good at, it was selling. That is what made IBM the giant it had become. IBM marketing was the best in the world and it ruled the company. Our sales were helped enormously by the fact that IBM was selling the systems to others **and** to themselves. ROLM switches were being purchased by virtually every IBM facility in the world. Actually I'm not completely sure that it was a worldwide thing (I suspect that it was) but I know for sure that they were being installed throughout the US and Canada. Why would they not want it everywhere? We made the best switch in the world! But as a result of IBM's decision to install a ROLM switch in each and every IBM facility, our requirements began to exceed our manufacturing capability.

The events that transpired in the months following the IBM purchase of ROLM had a tremendous influence on my way of thinking with respect to the value of employee involvement. IBM at that time was a very formal company with some very formal rules and practices. This was especially true for managers. Where ROLMans worked hard but liked to celebrate, IBM managers seemed to think that work was its own reward. Of course, the little department picnics and dreaded socials involving beer were discontinued. I think that it took about an hour after the takeover before they were gone. ROLMans for the most part were saddened with the loss of the picnics. Many of the employees looked at those little socials as a company benefit. Our celebrations were gone and we truly missed them.

The loss of celebrations had an effect on the culture of the plant. For years we had come to believe that every employee had a lot to do with the success that the company was realizing. The little picnics were looked at as recognition for jobs well done. Sure, we had a good product, but it took good, hard working people to get that product out the door. This was especially true with our product. As I said, it was very difficult to manufacture.

ROLM gave a lot back to its employees. Our motto: ROLM, A Great Place To Work, was not just words. We lived it. We spent a good amount of time celebrating something or another because it motivated us. It might surprise you to know that a lot of that time we spent enjoying ourselves was on company time. I would like to point out another, very important fact: we did not have a celebration unless we were caught up on our work. When there was work to do, we got it done. After all, this was not just a place to have fun. Work came first at ROLM but we felt that it was important to have a good time while we were at it.

IBM made a lot of changes at ROLM. They did not do so because they were trying to make life miserable for us. They had rules and they expected their rules to be followed. Initially, there were two companies with two different ways of approaching business. When one company purchased the other, the owner's rules had to prevail. Certain things had to change fast.

I think that when they found out that there was beer being consumed at some of our celebrations, a few of the IBMers had a cardiac arrest, even though we kept the drinking to a minimum and never came close to having an incident.

Looking back on it all, I think that they were right in their concern about drinking at a company function. It always made me nervous, that is for sure. There could have been serious problems as a result of it, although as I said, there never had been. If you think that it was the loss of company sponsored beer drinking that the employees missed, you would be wrong and you would be missing my point. We missed celebrating, period! IBM was very slow to celebrate anything. It is almost like whatever we were doing was expected of all IBMers. You worked hard, kept your nose clean and the company paid you well and gave you good benefits. ROLMans believed in all of that good stuff, we just thought that we should have a good time while we were at it. That kind of thinking is now a tenet of FCP: **Celebrate the victories as they happen!** Just do so in a sensible, yet positive way.

IBM brought to ROLM a more serious way of approaching work. It is not like IBMers were lousy people coming in bent on inhibiting fun and games. The IBMers we knew were also good people; they just had a different way of looking at things. It is not like they did anything wrong; it is more a case of them not including something. They certainly knew how to train their managers. About six months or so after the purchase, all of the ROLM managers were required to go to IBM management school, which actually turned out to be a great experience. The training lasted about a week and it was a first rate management training program.

Looking back on that training, I am convinced that it was the best management training in the world at the time. I learned a lot from it. The instructors were top notch and enthusiastic. The training was intense and required the trainees to focus on the task at hand not only during the day, but also continuing on into the evening. Every one of the ROLM students that I talked to was enthusiastic about it. The training that we received taught us that we did not just give orders; we had to help our people build careers. It also showed us that IBM was as interested in the worker as ROLM was. With IBM, the interest just presented itself differently – it was much more serious in a "Blue Suit" way.

Departmental celebrations, including picnics had been discontinued. Although this was bad news to many ROLMans, the net effect that the decision had on productivity is still open for interpretation. Before IBM, we made our shipments on time; afterwards, we still managed to make them on time. The difference was that all of a sudden we seemed to be working a lot of overtime. Before IBM, I almost always left work on time; afterwards, the majority of my workdays lasted at least 10 to 12 hours minimum. Quite often I was required to work weekends. Around the time that IBM sold ROLM to Siemens, I rarely had a weekend off. Is there a picture developing in your mind?

After all, were not we paid to work, not party? But then, if we were working more, why were not we putting out more switches than before? Why wasn't our quality better and why weren't our profits improving? Many of my employees began bringing up their displeasure over the amount of overtime that we were working. They were not just griping though, they were really upset over the fact that we just did not celebrate the **wins** any more, and they saw no end to the overtime. That cockiness that we used to have was replaced with a grim resolve. Work was just not fun anymore – not because we were working harder, but because we forgot how to enjoy work.

Even though we were saddled with more work than we could ever complete, orders we had no way of completing on time, and more WIP on the floor on a given day than we could finish in a week or two (our cycle time had gone up), we had access to, and some of us even got some serious CFM training. We never had the opportunity to really implement CFM throughout the factory, which I really wish would have happened. We did do a little experimenting with Kanbans but the experimenting did not last long.

I think that IBM management never really wanted to "push" CFM on us. We were too far west (or maybe not far enough to the east) and I think that they were trying to be sensitive to our "independence". After all, they had taken our name away, (which was later returned when they sold us to Siemens). Also, I think that they were trying to handle us with kid gloves. We were informal and they were very formal. They liked suits, and we did not. They did not believe in any kind of informal celebrations taking place during the workday and we certainly did. I look back on that time and come to the conclusion that they must have thought that they married a hippy.

As managers, we received a lot of training in just about everything. Many of us had the opportunity to attend seminars and actually try some of the stuff we learned. I really took advantage of the training opportunities, attending as many training sessions and seminars as I could on a variety of job related subjects. Many of the seminars were conducted away from the plant. IBM really tried to do what was right with ROLM; they just were not accustomed to our way of approaching work in a serious yet informal way.

Looking back on all of this, I sincerely believe and teach to this day that getting employees involved in process decisions and celebrating accomplishments with them is fundamental to success in production. Involving employees starts with making sure that they are armed with a good set of process values. Involving them requires that they be given responsibilities and that they be held accountable for those responsibilities. Management just needs to make sure that the changes are in line with acceptable process values.

IBM changed the way ROLM looked at meeting order commitment. Before IBM, the sales people processed orders with consideration towards meeting the order dates. Orders weren't accepted unless the date could be met. Under IBM rule, sales sold as much as they could and put the onus on manufacturing to meet the time line. In most cases, it was impossible to do so. As a result, the dreaded production meetings were held daily to make sure that the hot orders were given PRIORITY! I learned a lot from this practice – I learned that it is absolutely destructive to the manufacturing process and the bottom line. It hurts a company in every way that a company doesn't need to be hurt.

Managers, Bloody Managers!

In the ROLM world, celebrations always included recognizing and celebrating personal and team accomplishments. In the IBM world, recognition was mostly given to managers whose departments were doing well. When it came to recognition, it was usually the policy for managers to announce their department or company accomplishments and have meetings where **mostly** managers were recognized for improvements to their areas or to just cover some recent success stories. Workers were sometimes recognized formally, but not often. Managers usually waited until there was plenty to brag about to have a "recognition meeting" which discussed accomplishments that happened some time earlier. The ROLM celebrations were much more timely, and thus effective.

This is another way of saying that management did not celebrate often, and then usually only among managers. More often than not, the people on the production floor did not even know that their managers were out there bragging about them. Many times managers waited so long to recognize achievement that the recognition lost its significance. I have learned a great deal as a result of having experienced both the ROLM and IBM philosophy towards the recognition of accomplishments.

After managers celebrated among themselves, they usually showed their concern for the employees by having a series of all-employee meetings. IBM was very big on all-employee meeting. They let everyone know how successful the company had been lately thanks to their outstanding abilities as managers. Of course, everyone there knew that it was the workers who pulled off these miracles. It was true that management deserves a great deal of credit for its decisions that resulted in improvements, but, in my mind, involving and recognizing the workers for their cooperation and involvement was absolutely essential. Managers gave themselves bonuses – they gave the employees kudos.

I believe that managers should recognize their people first and then brag to their senior management about their own accomplishments afterwards. They should **constantly** celebrate accomplishments with the workers on the floor, in a timely manner, not just occasionally. Frequent celebrations send the message that employees' efforts really do matter and that the company just would not be successful without them. Employee involvement is just another buzzword that really is not understood by many in management. They might understand the definition; they just do not understand the proper application.

I do not recommend that management just walk around and give people the "Good job" cliché because they usually come off every time as being totally insincere. It sends out an unintended message: phony! Celebrating an improvement with the workers sends a different message: Sincere! Some managers spend too little time, and some spend too much time on the production floor. Each of those practices sends out a negative message. Walking around occasionally and asking pertinent questions while expressing sincere interest in the answers sends out a very positive message of concern. The beauty of this practice is that it does not have to be done that often.

There are ways to recognize employees that does not cost a ton of money. I have seen employee recognition programs that had a great deal of positive impact with the employees and the cost was minimal. First of all, the recognition was timely and wasn't designed as a popularity contest or a political victory. When the employees were recognized, they themselves really believed that the recognition was deserved. You cannot wait until the deed is forgotten because that sort of practice smacks of insincerity. Celebrate often and recognize quickly and watch your employees respond positively.

People out there know that most managers, especially senior managers, have more to do than just walk around and talk to workers. On the other hand, supervisors, should spend a at least two thirds of their time on the floor. That is not often the case in many companies with which I have worked. The reality is that supervisors spend about 90 % of their time in meetings. I have even seen lead technicians and key technical people spend three or more hours in meetings daily. In many cases they are representing their supervisor in the meeting because the supervisor is tied up with other meetings. Because these are people who gained their positions based on their knowledge of daily operations and their technical ability, their knowledge and experience is needed on the floor. Where are they? They are locked into one meeting after another. They are good at what they do, so we send them to meetings. What a waste!

Simulation Modeling

There was one person whose influence made me believe in simulation modeling as a powerful productivity tool. That one experience, which I will explain later, was so convincing and dramatic that it later would influence me to include simulation modeling as a key element of Fast Cycle Production. The person that I'm talking about was not one of the great innovators like Deming or Shingo, but an engineer just out of college.

When presenting recommendations for change to management, I found that just presenting the recommendations to them in a conventional manner didn't always work. I eventually discovered that simulation modeling, properly used, was able to do the convincing for me. Prior to that, I had no such magical tool. I had to use the power of persuasion and my own credibility in order to sell an idea. Once I found out what a powerful tool simulation modeling could be, I became an advocate of it and eventually made a mini-career out of teaching its use to others. The first time I used it successfully, I was hooked. Simulation modeling is now, and will continue to be a very important ingredient of Fast Cycle Production. FCP has a rather unique and effective way of teaching and using simulation modeling which is covered later.

Back in the 1980s, simulation-modeling programs were about as easy to learn and use as nuclear science. After I took a class on a General Purpose Simulation Software package (one of the earliest simulation programs out), I came away being impressed with the analytical (what-if) capability of the tool. However, I was overwhelmed with how difficult it was to learn. In other words, I considered it a great tool if only I could improve my IQ or if the software suddenly and magically became intuitive to me. I could see that if simulation modeling could be *conquered*, it could allow me to analyze a process and do what-if scenarios without disrupting the workplace. If I could master the program, I knew intuitively that the tool would be highly useful.

Another problem that simulation modeling posed back then is that it was virtually impossible to **validate** the models because they were so darned complicated and there was little or no error checking. (Validation: being able to say with greater than 99% probability that the model is a true reflection of the process being modeled) Even today, validation is an essential part of simulation modeling, but with the newer programs, it is much easier to validate the models. In my opinion, those early programs were absolutely impossible to validate. The other side of the coin was that when you created a model, there was no way that management could find something to challenge in it. They either accepted it or they did not. Simulation modeling had not yet become truly useful. It was in its infant stage and was not yet ready for prime time.

The first time that my interest in simulation was truly sparked was around 1984 when we were visited at ROLM Colorado Springs by a group of CFM people from IBM who came to see if a CFM implementation at ROLM was feasible. That is another way of saying that they wanted us to hire them to implement CFM in our plant. After all, they were an independent group trying to pay their way by picking up Blue (IBM to IBM) money. Because I was convinced by this time that CFM was the greatest thing since sliced bread, I was very interested in what they had to say at their presentation. After all, CFM was "The Americanization of JIT tools and values".

It was during that visit that one of the visitors showed me a simulation program that he had personally written. It was a simple program with absolutely no graphics or report generation capability. It was also very limited in what you could use it for. But even though it was limited in its flexibility and use, it was a simulation program. The main problem that it had was that it did not allow for stochastic representation of variables.

I'll never forget the guy that developed the program, because he was a member of the engineering elite of IBM, **and** he was assigned to the CFM team. That fact said volumes about IBM's commitment to CFM because they cared enough to put one of their best engineers on the team. The program he wrote was very un-complicated and easy to learn, but again, very limited. It was great for sizing Kanbans and little else, but it would be the first time that I had seen any kind of simulation modeling programs that were actually used by manufacturing people. It required a little knowledge of math, but beyond that, anyone could use it. I'm not even sure that it was really a simulation program in a strict definition of the term, but that is what he called it, and I accepted his definition.

This was simulation modeling that could be used by production people. It started me thinking about the possibility of simulation modeling programs that were powerful yet simple. I wondered why more companies were not developing better programs that did more, like for example, animated graphics that could show what the real process looked like and the movement of product through the line. Little did I know at the time, but that type of program would be developed soon and would have the features I dreamed of and more. Today, simulation modeling is used extensively in virtually every industry.

At the time, I was managing an in-circuit test department. The plant was fairly new, and the designers, in their ultimate wisdom had determined that my area would be a showcase for a very complicated conveyor system. The printed circuit boards (pcb's) would be loaded on the conveyor and transported to a test station where they were individually tested on standard pin contact testers. The test technician would direct the failures to analysis and repair stations via conveyor. The first time that I saw this process, I wondered why failures had to be transported anywhere. Why couldn't the repair station be right next to the tester?

The boards that tested "good" were sent along a different route to a point where they were removed from the conveyor and placed in a tote. The technician that was testing the boards had to make sure that every board followed the proper route. Someone else had to make sure that a box of fifteen identical pcb's somehow were accumulated later and did not get mixed up with other product. Again, I wondered why the boards had to be transported at all. Why, I thought, couldn't they just be accumulated at the tester? The boards spent more time traveling around than they spent being tested. You'd think that we were in the business of transporting product, not testing it. I guess the designers liked to see the product constantly being moved.

If you can imagine in your mind what kind of conveyor system this was and how much space it took you are probably conjuring up visions of horror. If you're not, you are a manufacturer of conveyor systems. At this point I would like to point out that there are excellent applications for conveyor systems. This was just not one of those applications that made sense. This system made our work life a living hell. It was difficult to maintain and wasted a lot of space, not to mention the fact that it gave us a lot of maintenance problems. After we got rid of this beast, we were able to combine the test department and the manual assembly line into the same space originally assigned to in-circuit test The conveyor system was not an enhancer to productivity, it made testing and repairing those printed circuit boards much more difficult than it should have been.

My department was always behind schedule because we spent too much time **managing WIP and that darned conveyor!** The person testing had to wait until the whole line was purged before he or she could start a new part number. This cost us a lot of machine utilization time and output. There was no way that we could keep up with the demand requirements because we just did not spend enough time testing. We were the bottleneck. The conveyor system was unbelievably cumbersome and wasteful. It required too many people to staff the line and it adversely affected quality. I never found out who designed that monster because nobody would admit to having any part of it. My boss thought that some **consultants** had designed it while they were building the plant. He thought the consultants were from the company that made the conveyor. That sure soured me on consultants (until I became one) for a while. I surmised that our competitors were paying the person(s) who recommended that system.

I was determined to convince my boss and the plant manager that this white elephant should be replaced by a process that would take up much less space and would improve productivity. Imagine the concept: put tester, analysis and repair within an arms length of each other (see Single Unit Flow, Part 2, Chapter 2). There would be no transport of product until the boards left the department for good. I was sure that I would be able to sell my idea easily. Anyone in their right mind would agree with me, I thought.

In those days, conveyors were very popular, but in many cases were very inflexible and difficult to modify when the original design was not appropriate for a new product line. The plant manager had been convinced that this system was a good one and had approved the money for it. I couldn't believe that he was sticking up for something as worthless as pond scum. Here I was telling him that he had made a mistake and that he had invested in a piece of manure and he was telling me that I should make it work more efficiently. What it all really boiled down to was that this monstrosity had cost almost a million bucks, and to write it off would be too expensive in the opinion of the plant manager.

I had to prove that it was in the best interest of our bottom line to remove the beast. I had to make a business case showing that it would be cost effective to remove it and replace it with a more productive process. My first attempt on the business case was to calculate manually (spreadsheet and graphs) what the results would be if the recommended changes to the line were made. I had to include the cost of the removal of the conveyor and the cost of any new workstations, and present a design for the alternative process. I drew out a new layout showing the new locations of the workstations and made a presentation to top management. The plant manager decided to involve all of the department heads in the final decision to approve or reject my recommendations.

The senior staff rejected my proposal, but they were (sort of) positive in their rejection, if that is possible. They said that the numbers sounded pretty good, but they were unconvinced that I could achieve them. They trusted me because I had a good track record for being able to do what I said that I could do, but still there was doubt. They just couldn't take the chance. With so much at stake, they couldn't take the chance that I would be proven wrong later. Perhaps they also thought that somewhere out there, was a manager who could make the old system work. In any case, they did not close the door on the idea; they just told me that they remained unconvinced. When I asked what it would take to convince them, they told me that I just had to come up with proof that the new process could perform as I said it could. The battle lines were drawn.

Using simulation to make my case just never came into my mind. First of all, I was not very good at working the programs that I had previously been trained in, and second I still was not convinced that simulation would help my cause. There was nobody around that could use the software any better than I could at the time. Besides, I had never really seen a simulation model that did anything for anyone. I still have the textbook in my library. Every once in a while I look at it to see if it looks any simpler now that I am more knowledgeable?? Yeah, you're right, it is still too hard to understand.

I was searching for a way to convince my management and nothing seemed to be working. Simulation was not even under consideration by me – yet. The saga continued. I remained deeply convinced that this monster that I was saddled with was as worthless as spit, but could convince nobody (with authority) that it should be dismantled and replaced. It looked like a hopeless situation for me.

Suddenly, I got a break. Out of the blue, an engineer assigned to give engineering support to my area approached me on the issue of my business case. After one of my presentations, the engineering manager (her boss) had come away convinced that my proposal had merit. He showed my presentation to his staff and asked for their input. One of his reports, a young woman that had recently graduated from college told him that she had taken a simulation modeling course just prior to graduating and still had the software. She suggested that maybe the potential gains could be quantified and validated by simulating the baseline, (as-is) and comparing the output data with my proposed changes, using the simulation software. The program could quantify the results in output and dollars and provide us with printed reports. He freed her up to work with me and she wasted no time. She volunteered to create a model for me and I accepted her help with pleasure.

We worked together to gather the data for the model. As I said, our first goal was to create a baseline model that accurately represented the process we were modeling in both output and resources. We did so successfully while meticulously documenting our model in the very same way I would later adopt. Once we were convinced that we had a valid model that mirrored the current process, (the baseline) we presented the results of the original model to management, answered their questions and ended up convincing them that we had effectively duplicated the process.

In other words, the conveyor system in the model performed identically to the existing conveyor process. The testing in the model took exactly as long as the testing in the existing process. The machine downtime in the model was identical to downtime in the existing process over time. Our hourly, daily and weekly output in the model was identical to what we had been doing over time with the same amount of people. We had recreated the test process in the simulation model and the output reports backed that up. Bottom line, we convinced management that our model was valid. The important first step was completed.

Next, we created the what-if scenario that showed how the process would behave when the changes that we recommended were implemented. The layout that I recommended was created in the program and all of the time and motion data was adjusted appropriately. We made sure that the scenario was fully documented and all of the dependencies were explainable. My accomplice did all of the simulation modeling and she did a tremendous job with it. We had put together a very convincing business case.

When we did the presentation of the models results, it was not even a contest. The results of the model showed them that it was in their best business interest to approve the proposed changes and write off the conveyor system. We gave them payback time, resource needs, output changes in both product and dollars. We gave them the works! The result of all of this was that we came out of that meeting with a unanimous approval for everything we recommended.

The rest, as they say, is history. The simulation model brought credibility to my proposals because the results shown were even more dramatic than I had anticipated. The results showed that not only would the output capability improve; it would improve by 60 percent. Imagine being able to increase your output capability by 60 percent, and quickly! Equally dramatic was the fact that I could get this output increase with 40 percent less people. I want to make sure that you got this: **More output – Less people!** How often do you see that?

The cost of removing the equipment had a one-month return on investment. The cycle time of the product had been reduced almost ridiculously. All of the metrics that were important to management had been improved more than anyone, including myself had anticipated.

The changes were made and the results were just what the program predicted and more. Our backlog was eliminated within a week and we were no longer the plant bottleneck. Let me tell you that a big load came off our back. The workers liked the new layout so much that they made it work better than we hoped because the new layout made it possible for us to develop integrated work teams, where everyone was accomplished at all of the different jobs and were able to rotate, rather than have Testers only testing and Rework people only doing rework. More on this when I cover the **Integrated Work Teams** and **Single Unit Flow** (Part 2, Chapter 2). Our department became pretty famous for its turnaround. I received a bonus and a great deal of attention and that young engineer got a well-deserved promotion for her contribution to the project.

I had discovered the **real** power of simulation, which was its credibility with management. I had known all along that the proposal would work; all I had to do was find some way of proving it without risk. Simulation was the tool that I used to present my business case. All we had to do was show that the baseline model performed identically to the real process in terms of critical metrics like output and cycle time. Then, we made the necessary changes to the baseline to create the "new" process, then ran the model for an identical amount of time (one year), presenting an apple – apple comparison. The program produced reports proving the new process with the proposed changes was clearly superior to the original process, something we knew all along. We did not have to give a brilliant, silver tongue presentation to convince management that we had a good idea. The software was plenty convincing enough. All we had to do was to convince management of the credibility of the program and the models that we created. A quest that had begun almost a year before had ended abruptly and positively by just using simulation to make the business case. I had found a powerful new friend!

From that day forward I was a believer in simulation modeling, but would not use it again in a project for several years (When IBM sold ROLM). Simulation was still thought of as a tool that was only useful to engineers to evaluate machines and tools, not as a tool to be used by IE's and manufacturing people. By the way, my young collaborator was an industrial engineer and simulation modeling was not a requirement in her curriculum, but an elective. It was not yet a viable manufacturing tool, because it was just too complicated for people without an engineering background to use. It was not that it required a superior intellect; it was more like you either understood it intuitively, or your education prepared you for it, or both. Even today, with simulation software that is easier to use and learn, I find that some people take to it and some do not. If you take four people with identical degrees and identical IQ's, you might get one or two of them that is really good at simulation.

In the beginning (So to speak) there was too much programming knowledge required in simulation programs. Mere mortals (without an engineering background) were not equipped to deal with it. Those early simulation programs were nothing more than very high-level programming languages. If you were not skilled in programming, you had better leave them alone. Also, they were not graphically oriented. The early programs did not give you a picture that **looks like** the real McCoy. You did not **see** little people running around the screen, and you did not see machines that looked like the ones in your plant like the programs that are available now.

Chapter 6 – Consulting: My New Career

Because of its complexity, simulation modeling was primarily an engineering tool. That scenario would change dramatically in the late 1980's when the makers of Witness™ and ProModel™ would develop programs designed for manufacturers that actually used manufacturing terms. These programs allowed one to create a realistic graphic presentation of any process. We're not just talking graphics here; we're talking animation and program data showing (and automatically updating) on the screen while the program runs. It was still necessary for the user to understand statistical probability and even the use of statistical distributions, but even the statistics were simplified by the program. This is a real plus when you are attempting to verify that the model is behaving as it should, and it is essential in validating the model. I did not know about these new simulation programs until later. Simulation modeling would be the catalyst that would launch my consulting career.

The opportunity came when IBM made the decision to sell ROLM. As I've already stated, I had helped ROLM Colorado Springs to improve its own processes and my accomplishments had not gone unnoticed. A manager at IBM Boulder that was starting up an external consulting team had somehow become aware of my productivity accomplishments and set up an interview with me. He was looking to put together a team of people that would work with clients outside of IBM, as opposed to the old "internal" consulting policy that had been followed in the past. After the interview, he made me an offer to move to Boulder to join his team and I accepted. It turned out to be one of the best career decisions I had ever made.

Eventually, this team would turn into the very successful IBM Consulting Group. There were pilot teams like ours being formed at a few other plants. These teams would eventually form a very formidable group of people that would help to make IBM Consulting a force in the consulting world. At that time they called us the "Design Team", for reasons that I still do not understand. Perhaps this team was to form the "design" of the consulting group, who knows? At any rate, I moved to Boulder and my career as a consultant had begun.

When I arrived at Boulder, I quickly discovered that I was just one person in a group of individuals trying to find a niche. We had to develop skills that were needed by other companies. Let me rephrase that to say that we were trying to develop skills that clients would be willing to pay for. We had to have something that we could sell to other companies that would help them to improve. We couldn't just show up at a client's door and say, "I'm an IBM consultant and I'm very knowledgeable - Hire Me! In the world that we were entering, you had to be able to convince the client that you were worth the investment that they would make in you. I would learn very quickly that there are a lot of people that call themselves consultants, as there are a lot of people that call themselves salesmen and saleswomen. There are not nearly as many of them that can be called successful

Being a knowledgeable manufacturing man was not enough to get work for myself. I had to come up with something else, something with more substance that would attract clients. There were already a couple of guys on the team that were doing simulation modeling. They pretty much just worked in their office and did not go out on contracts - much. Truthfully, at that time, there were very few Design Team members that were going out on jobs. Most of them were just trying to find something to do and develop a skill that could sell. Many of them never did. As I would discover, consulting is not for everyone. Like any successful salesman, you first have to have something to sell, and you really have to believe in what you are selling, Just having the ability to give a good presentation is not enough to motivate a budget conscious executive. A silver tongue won't get the contract when substance is required.

My mission was very clear because I was determined to be successful. I had to develop a consulting niche that would sell. This meant that I had to have something that could do good things for a company's bottom line or relieve them of a problem that was hurting them. I already had one thing going for me: I had a manufacturing philosophy that I really believed in (JIT and CFM), and I was not shy about talking about it. You could probably change that statement to read: I was opinionated, stubborn and looking for a good debate. But I needed more; I needed a **methodology.** I had to be able to lead the client away from where they were and take them where they wanted to go.

Since the Design Team had already purchased several copies of some new simulation software, which was being used by several Design Team members, I naturally had access to it. When I was introduced to this new simulation software, (Witness™) I was blown away by its simplicity of use and its power. Witness™ had everything going for it: It was powerful, had pretty good graphics and statistical capability and allowed the user to customize everything from the display to the reports. It still required that you enter some logic to get it to behave under certain circumstances, but I had some programming training so I was not intimidated by it. I found the conventions it used easy to understand and get used to. I decided that I would sell simulation modeling consulting. It should be noted that Witness™ and ProModel™, another fine program, have improved dramatically since those early days, but the earlier programs were outstanding even in the early 90's.

Initially, we in the Design Team did not go out and sell ourselves to clients; we had IBM marketing sell us. We had no idea how to get an audience with a potential client. Heck, in many cases, the marketing people did not even know what we were selling. Since marketing traditionally sold only hardware, and not services, the situation was delicate for both of us to say the least. They first had to become aware of us and then they had to find a fit with their customers and us. Since they did not know what they were dealing with, they were slow to push us to their customers at first. We were at the mercy of marketing and they held the trump card. Compared to what they got for selling a mainframe, what they got for selling us was peanuts. But if their customer decided they wanted us, and we did a good job, the potential for the marketing person to enhance their position with the customer was huge. Alternatively, we also presented a risk to their relationship with their client.

The whole relationship between marketing and the Design Team began to develop when they saw us as a way to improve their image with their customers. If the marketing person could see a customer need for a service that the Design Team provided, they would suggest to the customer that they allow us to come in and look things over. If the customer agreed, the marketing person would contact our office and arrange a visit to the customer location. Once there, it was up to us to look things over and make a presentation of our findings. If the customer was impressed enough to enter into a contract with us, we were in business. In those early days, very few visits turned into contracts. Fortunately for me, I was an exception.

Since my skills were mainly in the area of manufacturing, I was tagged as a manufacturing expert (Duh). Unfortunately, just about everyone on the design team was a manufacturing expert so that dog didn't hunt. I had to be different, so to speak. I had to have something more than that. I decided to become a manufacturing expert that taught simulation modeling to clients. I talked my manager into giving me one of the Witness™ security devices to use and I was in business – almost. First I had to learn how to use the software.

I attended a training class on Witness™ and practiced my newfound skill by creating little models in my office. I used my old line as the process I modeled for practice. I didn't forget the manufacturing values that I had acquired, but rather added them to my training methodology. I then formalized and documented the simulation methodology that we used to create the model at ROLM. I developed the plan for documenting that I still use today, which was a spreadsheet presentation of all of the critical data included in the model, which I named the "Data Matrix". I now had a product that I could sell to others, namely simulation modeling training. I would teach people how to improve their processes with simulation modeling. I take that back. I would teach people how to **use** simulation modeling to improve their processes. But then, how would they know what was wrong with their processes? How would they know what the solution was? Yes, Virginia, the truth is coming. Now, back to the story …

About 2 months after my Witness™ training, I had the opportunity to join a group that was visiting a company in Spokane that made, among other things, computer keyboards. This particular company had developed a brand new product and a new process that had some robotics in it. They were very proud of their new keyboard, and rightfully so: it was using a new kind of resistance under the keys instead of springs like other keyboards. The device that replaced the springs offered resistance to key press that was consistent and had the feel of the springs. This new keyboard could by made easier and more economically and they planned on selling a bunch of them. They wanted to make sure that they hired the right amount of people and machines to meet the projected demands predicted by their sales people.

We were given a tour of their factory and the new keyboard line. When we went to the meeting room to tell them what we had observed and what we thought of it, we only had our notes to go by. As we were walking around, I took notice of the WIP buildups at several points and saw some things being done that went against my grain. It was not too long before I was convinced that they really were not very productive. In our team caucus I was told not to come off as being negative, but to present my observations in a way that spoke of opportunity, not wrongdoing.

I followed the advice and presented what I had seen. There was a older gentleman in the back of the room that was very interested in what I was saying. He asked some very good questions and challenged some of my reasoning. I answered his questions and explained why I thought certain things could be improved and what I would do about them. When I said that simulation could be used to quantify the effects of the changes I was talking about, the old guy really got interested. I didn't know it at the time, but he was the owner of the company. I had established my credibility with him. He told me later that he had heard of simulation modeling and was very interested in finding out what it could do for him. Until then, he hadn't found anyone that he thought could teach it to his people. He figured that I could do that for him.

I worked to convince the rest of the audience that simulation modeling was the best way to learn the answers to their questions. Although I was speaking to the audience, the old gentleman was really the only person I needed to convince. He got up during the meeting and said that he was sold on simulation. He almost hired me on the spot, even surprising the marketing representative that had brought us in by asking him if he had something that he wanted signed, like a contract. Since it was not expected that this would be anything more than a visit, neither the marketing rep nor any of us were prepared to make an offer. We suggested that we put together a proposal and return in two weeks. The suggestion was accepted. We went back to Boulder and I put together a proposal for a contract – my first. It wasn't long before we had a signed contract – my very first contract! My new career had truly begun.

That was my first contract, and it was a huge success. I trained 2 of their engineers in simulation modeling and JIT/CFM values. Part of the contract stated that the company would purchase its own simulation software and get syntax training at the Witness™ training facility.

My responsibility was to teach them how to effectively use simulation. It would have been tough to teach them the software conventions, since I was not yet qualified for that task. I had received the same training that my trainees were about to receive just two months prior to this contract. As they say in the business, the teacher stayed about one chapter ahead of the student. The difference between my trainees and me was that I was bringing something besides simulation knowledge to the table; my students would learn how to evaluate their processes by looking for waste and comparing their existing process with waste free alternatives. Simulation would allow us to do the comparisons without altering the process.

In the two weeks prior to the contract, I tried to get as much practice as I could, while calling for technical assistance when I got stuck. I was quite aware that as far as the software went, I was barely one step ahead of my trainees, but I was miles ahead when it came to knowing how to identify opportunity and what to do when opportunity was recognized. Not only that, I had developed a methodology to lead my students through the simulation project (thanks to the conveyor experience at ROLM). Finally, it was not necessary that I be an expert in making keyboards because the opportunities that I saw had nothing to do with the technicalities of their product, it was all about the process. Being an objective outsider increased my value as a project leader because I was looking at the process with "untainted" eyes.

The first thing I taught them was how to recognize right from wrong, good from evil as seen through my eyes. I gave them a mini course in the JIT and CFM values that I had learned. That was the first time that I actually taught anyone the process values I had learned through my studies of JIT and CFM. I was pleasantly surprised how quickly and enthusiastically these values were accepted. They learned quickly and we accomplished a great deal during the training and analysis period.

Since I was the trainer in this project I had to learn the intricate details of the process. I had been given a tour of the line and I had a good idea how things went, but I needed to know exactly how the keyboards moved through the line. I had learned how to flowchart processes back at ROLM and put the training to work. The difference in this case was that I wanted to take the flowcharting to very low (detailed) level, a decision that turned out very well.

I decided that the best way to learn all of the details needed to create the flowcharts and the model was through interviews with the workers – another decision that worked so well that it is still being used in FCP. It was suggested earlier by their engineering folks that we use the flowcharts that had already been developed, but happily I resisted that notion. I'm glad I did because the detailed flowcharting that we accomplished, not only taught me the details of the keyboard manufacturing line, it brought to light all of the problems that we would later exploit. Just gathering the data taught the team more about the process than most of the people that had worked there for years knew.

I also discovered with that first experiment that not only did **I** learn a lot, my students also benefited from the experience. Flowcharting their own process taught them some things that surprised them, and they worked there every day! As I said, it was not only educational; it helped us to arrive at meaningful solutions because we were looking at each step of the process in a critical manner. From that day forward, I used process mapping in every one of my projects. It is still a permanent element of FCP today.

Throughout the project, we would go over the flowcharts and update them when we discovered an error or noted a change. Flowcharting had become part of my methodology through unforeseen necessity. From that point on, even though I continued to use process mapping, I would also continue to modify the process mapping methodology to make it even more useful (Part 2, Chapter 3).

I learned that you can never "fix" something unless you really truly understand it. Process mapping helps you to understand the most minute details of a process. Process Mapping will lead you to solutions that you never would have found without it. That is, when you do it right, which is saying a lot. This scenario does not just apply to manufacturing issues; Process Mapping is very much applicable in non-manufacturing situations. I know this to be true because every time that I work with service or support organizations such as engineering, maintenance and non-manufacturing companies such as banks, process mapping is the tool that consistently pays dividends for the client.

Back to my first project …starting with the flowcharting of the process, we went on to create an accurate baseline representation of the "as-is" process (just like ROLM). By analyzing and applying sound values to finding the solutions to the problems we found on the line through process mapping and simulation, we came up with some great solutions.

We were able to find and open up the bottleneck by eliminating several redundant tasks on a machine that increased output just enough to eliminate the necessity of buying another, very expensive machine that was already on order. The main point here is that we opened up the bottleneck by eliminating unnecessary process steps, thus enabling the process to produce more keyboards each day. We found the bottleneck; found out what could make it produce more and implemented the solutions successfully. Simple but beautiful!

We did an excellent job on this project. If the client had known that I was feeling my way through the project barely one step ahead of the students, it would have caused him to be concerned. But although my "project methodology" was a little ad-hoc, the values that I brought to the table were rock solid. We found solutions that fit our value set. Consensus was easy because we were all looking through a single set of eyes. That simple fact has carried me through a number of implementations: I made sure that the team that I worked with saw opportunity and solution through a single set of eyes - mine.

What did we accomplish? Well the client was more than happy with our recommendations. In our presentation to management, we showed how they could cancel the order on the machine that they had planned on buying "just in case," saving approximately $100K. We also showed how they could meet their 3-year projections in sales with just one shift, instead of two.

They had already anticipated starting up another shift and were in the process of interviewing candidates for second shift. Since the bottleneck was opened up, they could meet their projected output demands without adding people. This discovery alone saved the company tens of thousands of dollars in salary and benefits monthly. This discovery alone paid for the project several times over.

The cycle time of the product was so quick with the process changes that we recommended that one shift was more than enough to meet projected demands. The amount of WIP that was on the floor was reduced by more than 50% as a result of our recommendations, and one inspector was eliminated because there was no need for inspection at that point in the process. The ratio of output per person was increased by 25% for a given period of time. Bottom line, all of the measurements critical to profit and productivity were improved.

I learned something else from that project that did not really sink in at the time. I did not recognize it because this was my first consulting project and I had no basis for comparison. Later, this critical issue really sank in because I was to learn that all top managers are not created equally when it came to resolve. Each of the recommendations were quickly accepted and implemented. Let me emphasize the critical word in the last sentence: QUICKLY. Why did it happen quickly? **It happened quickly because the owner made sure of it. He pushed his people to get the improvements implemented, and he simply did not let up on his demands.**

Our biggest advocate was the big boss - the owner of the company. It was obvious to those that worked there that when he told people to do something, it was in their best interests to do it because he always followed up on his requests. This might seem a bit authoritarian to some, but my point is that when you have executive support in a project and they are not only vocal, but also active in their support, things happen, and they happen quickly. Not only that, the improvements stick when there is continued top level focus on them.

The lesson would eventually become clear to me: When top level managers just hand off projects and then walk away from them, while "assuming" that everything is happening as it should, they will one day find out that things are not going right at all. I'm not saying that they have to constantly monitor the actions of those they delegated the responsibility to, I'm saying that they have to make sure that people understand just what is, and what is not acceptable. They have to set the standard for maintaining the value set that they want in place and take action to make sure that there is no erosion of the values and their application.

My first project had those ingredients and the success was profound and lasting. What was missing? We had not put in place a process of continuous improvement. There has to be something that ensures that not only the original solutions are implemented, but also that the effort to continuously improve is sustained indefinitely. TAKT Management is that system, but unfortunately it was not born yet. Simulation by itself only helped to find and evaluate solutions to problems that existed when the project first started. As I've stated, my methodology was still in its infant stages, but I was learning something new every day and making modifications to my methodology.

My new career had begun, and for the next three years I did very well at selling simulation to companies in need of improvement. Throughout that period, I was 90% billable. I still did not just teach simulation to my clients, I continued to send them to the software company for training, but I taught them how to get the most from simulation and I continued to fine-tune my methodology for simulation and project management.

I taught my students that simulation was just a tool that usually proved what we already knew, just as I knew the solutions to my line back at ROLM would work before the simulation project began. There would be times that simulation brought surprises, but usually the surprises were in terms of a solution giving greater results than anticipated. The solutions were not hard to find when there was a good process that could seek the problems out and test the solutions.

Quantifying the results in advance without disrupting the process was the real power of simulation. Simulation quantified the effects of the changes in a way that management appreciated and respected. It reduced their fear of gambling on an investment in change by making them feel warm and fuzzy about the probability of success. Simulation gave instant credibility to the solutions with management.

I was having the time of my life! I could walk through a line and see large WIP buildups or something else awry and knew what to do about it. I did not just see problems; I saw solutions. My values were JIT and CFM based and therefore the solutions came from the same sources. What a piece of cake! It was as easy as giving away gold. All I had to do was to make my trainees understand the difference between what is good manufacturing practice and what is bad, and then allow their new values to lead them to a solution. I actually led my trainees to the source of the problems and they did the rest. What could be easier - right?

Well, as I've said before, all top managers are not created equally. Sometimes no matter how solid the proposed changes are, there are those that just do not follow up on making sure that they happen. Also, some top managers have far too much trust in the people that they put in charge of implementing the changes. The top manager fails to maintain focus, and someone lets them down. The failure, however is with the manager that failed to make sure that everything went as planned, not just with the people assigned to implement the changes.

There would be times some of my trainees would exhibit passion and fire to do what is right and then later wilt under political pressure or ambition. Most of those that I've trained have remained true to their beliefs, but some, faced with opposition from people in higher positions that believed otherwise, have found themselves going over to what I consider the "Dark Side", where political compromise is the norm. There is nothing that I, or anyone, can do about that. The really true believers are the ones that have not only accomplished a great deal for their processes; they have done well with their careers also. In their case, they chose to do what is right, not what is convenient.

Early in my consulting career I found that my trainees looked to me to be the one that stood up to anyone that challenged our recommendations or were resisting our efforts. I was the one that provided them with consistent leadership, in that I never waffled when challenged. No matter what the challenge, they could rely on me to stand firm and not compromise my values. I have always been true to my values because I believe in them so strongly that there is not really any room for compromise. If a solution cannot be found within my value set, I just have to do some more work to find one that does. I have found that **it is always possible to find a solution that fits within my value set.**

It really did not matter if I had little or no experience with my client's product. I did not have to be an expert in their field with a long history of experience working with companies like them. All processes are essentially the same and all of the solutions are fairly identical. I could rightfully say that I had seen their problems before because I had. The place was different and the product was new, but I had seen **all of their problems** before I got there. There was nothing new in any of them. They all had the same kind of problems that I had seen before; they were just presented in different forms.

In project after project, the presentations of solutions to management were those that cost very little or nothing to implement, had a great deal of impact once implemented and were easy to implement. Let's recap that statement. I made 3 important points about how to evaluate solutions. It is not necessary to hit the extreme end of each point, but each of them is crucial in the evaluation of solutions. Solutions should meet the following criteria:

1. **Ease** of implementation in terms of
    a. Layout restrictions
    b. Training
    c. Time required to implement
2. **Cost** of implementation in terms of
    a. Dollars
    b. Internal resources
3. **Payback** after implementation in terms of it is effect on:
    a. Quality
    b. Cycle time reduction
    c. Output

I still teach this method of solution measurement today. There is another element of solution evaluation that I use today, and that is called Manufacturing Cycle Efficiency, or Production Cycle Efficiency in non-manufacturing companies (MCE and PCE are the ratio of value add to total cycle time) which I will discuss in detail later, (Part 2, Chapter 2) but for now it is important to note that a solution that has a high impact on critical production metrics cannot be excessively expensive and difficult to implement. It should be an improvement that works, which is another way of saying that cycle time has been improved **and** the ratio of value add to total cycle time has been improved. If the solution has resulted in improved cycle time, it is not an acceptable solution unless the MCE has also been improved. They are tandem measurements in process efficiency improvement. I consider MCE to be critical, not as a standalone measurement, but one that is used in tandem with Cycle Time.

Throughout my career, I have seen company after company throw money and equipment at their problems in an attempt to improve their ability to produce. On each occasion it was determined after the process improvement project that the company had not acted out of true necessity. Some of the equipment that they thought was needed was not needed at all. In fact, we often found that the equipment that should have been purchased was not included in the budget. In many cases, it was found that the present equipment was adequate if only it was used properly.

Too often, companies attempt to improve their productivity and output potential through automation and technology before they attempt to clean up the waste in their process. This approach not only fails in most cases, the failure is very costly. Each automation effort should be preceded with a simplification effort. As I've stated, companies often take the opposite route: purchase the equipment and then try to clean up the process. Sometimes they just think it is enough to just buy the equipment with the hope that everything will work itself out. It does not, and the experience is sometimes as hard on careers as it is on the company pocketbook.

Going back to my simulation modeling consulting practice, I had created a methodology for analyzing a process to the finest level of detail and creating a computer version of that process that identified and quantified the effects of changes to a process. My students learned how to go all the way from analysis to meaningful change in a repeatable fashion. I taught my students how to create a simulation model of the baseline (As-Is) of the process. They learned how to collect and arrange the data for the model and to verify all of their findings. In addition, when the model is run, the reports from the model accurately represent the results of the original process (validation). Furthermore, all of the assumptions made in the creation of the model, such as statistical distributions are documented accurately. The methodology works every time.

Because many of my students were not familiar with statistics and statistical distributions it was necessary to teach some of them the basics of statistical distributions and why some distributions are more appropriate in certain simulation scenarios than others. It is important for them to be able to select the proper distribution for a certain stochastic representation such as, the time that it takes to perform a setup on a machine, or the time it takes to perform maintenance on another machine. Selecting the wrong distribution could easily invalidate a model. When there is reliable recorded data, such as frequency and length of machine breakdowns, a distribution can be created with empirical data and entered into the model. If decisions are made based on invalid model output data, the credibility of simulation itself could be seriously compromised. The results could be disastrous to profit and careers.

Once a valid baseline model had been created that accurately represents the process being modeled, the team was taught how to brainstorm, select and prioritize the solution scenarios. This they did as a team and it always seemed to me that they enjoyed this part of the project the most. There was enthusiastic debate on every solution as they made their case for promoting their own individual selections for improvement. From the start, I insisted that we as a team must come to consensus. By this time in the project, the team knew right from wrong, process wise. For this reason, I usually took a back seat in the discussion, guiding and facilitating, rather than directing and dominating. The team might ask questions, but they were in charge of the process at that point.

When the scenario models were developed, the teams really got excited about testing their solutions by running each scenario to test the impact of the changes. Each scenario was compared with the baseline model. In each scenario, Output, Cycle Time and MCE were compared with the baseline. There might be more metrics added, such as dollars saved, resource costs and other measurements that the team determined was critical, but those three critical measurements were ALWAYS given priority. If output, cycle time and MCE were improved, any other measurement, including quality would be improved also.

My simulation methodology worked very well for me and was hugely successful in every single project. As a result, I stayed busy teaching companies how to use simulation effectively. After my first contract with only two trainees, I recommended to subsequent clients that they assign four people to the project. I consider four people the perfect team size for a simulation project, and to this day I still prefer four or five people for Fast Cycle Production project core teams. I have worked with as many as eight successfully, but find it very difficult to give each individual the high level of training and involvement that they need to function effectively in a core team. Two or three people are not enough, in my opinion, because there is so much work to do collecting data, forming the model(s), and conducting interviews. The team would be stretched dangerously thin if one out of two or three is missing and heaven help us if both of them were absent. Four or five is a much safer number. In addition, as I have stated, not everyone really "gets it" when it comes to simulation modeling, no matter what his or her level of intellect is.

Some people understand it intuitively and others really struggle with it.

Being a simulation consultant taught me that my success at ROLM with simulation was not just a fluke. I found that most senior managers are quick to give their approval to process changes that they may have rejected without simulation. Simulation helps the proposed changes gain credibility with management! The best part is that the solutions that we presented in project after project did not just improve things a little; they had a **profound** affect on output, cycle time and MCE. I'll say it a little different so that you get the message – **once the solutions were implemented, management saw dramatic results every single time!** This means that they got their investment back quickly and made tons of money by approving the changes and seeing to it that they were implemented! The solutions presented had those three elements I spoke of earlier that determined the worthiness of each solution:

1. **Ease** of implementation
2. **Cost** of implementation
3. **Payback** after implementation

When a solution scores high in those three areas, you definitely have a winner! The solutions that qualify as "low hanging fruit" are easy to implement, have a low implementation cost and have a huge payback once implemented. Those are the solutions that you want to implement quickly. In many cases, you will have paid the company back and then some for their investment in the project. Many of these can be implemented while the project is still underway, which gives the project impetus. After the low hanging fruit is harvested, more complex solutions may take longer to implement, have a higher cost involved, but still have an outstanding payback once implemented. It is important to prioritize the implementation of the solutions wisely to ensure that the project pays for itself as it goes along. It is really not that difficult to do, it just takes resolve.

Simulation is not only an excellent way to calculate and predict the effect that the implementation of the solutions will have on most of the variables mentioned, it is also a very accurate way of calculating labor costs, value of product shipped and any other metric that affects the bottom line when the process is improved. In addition, it is possible to evaluate the effects that the changes will have on other parts of the factory. It is very possible to improve one department and create serious problems for another department or work center. Certain work centers may be unable to supply the improved department adequately or they may be faced with too much work at their input point. In either case, it would be necessary to deal with those situations to produce a smooth flow of product throughout the factory.

By measuring the *potential* impact of the solution, the risk of approving the solutions is reduced significantly for management. It is one thing to find something wrong with a process and make recommendations, but it is quite another for management to understand the risk, along with the cost and payback if the solutions are approved. If management is confident that the critical variables are accurate, they are much more likely to give their approval. Managers have very strict budgets to consider and they do not like gambling with those budgets. They want to be very confident that their investments will produce results and produce them quickly or they might just balk at giving their approval. If the simulation project is conducted properly, they will have the assurances that they need.

So, why was simulation so successful for me? Was it that I had discovered a tool that was revolutionary and risk free? Hardly! Obviously, if I had concluded back then that simulation was the reason for the successful projects, I would have been sadly mistaken. Certainly, simulation was a valuable ally in convincing management and quantifying results, but it was not the simulation tool that caused the improvements; it was the process values that were used to arrive at the solutions. I taught people to know the difference between what is right and what is wrong with their processes. I taught them how to detect the opportunities and identify solutions to those opportunities. Of course, if a bunch of meaningless models were created and the results were confusing and contained the wrong variables, the projects would have been meaningless. But arriving at meaningful solutions and being able to prove that they were valuable and credible was what simulation modeling was all about.

Chapter 7 – Forming The Methodology

My knowledge of JIT and CFM formed the foundation of my beliefs. From JIT I learned that there are two technologies by which you can effect change. One technology is process simplification and the other is mechanization or automation. It is through the proper combination of these two technologies that companies are able to achieve world-class status as producers. If a company uses only one of these two technologies, they will eventually reach a point of diminishing return, or will have to go back and cover ground that should have been covered earlier.

In my studies of his published works, I have come to the conclusion that Shingeo Shingo has strongly inferred that American manufacturing people just didn't "get it" when it came to the proper application of the two technologies: Automation and Simplification, and my own experiences have proven to my satisfaction that he was absolutely correct. American producers have traditionally shown the tendency to automate their processes **before** simplifying and eliminating waste from them. Many companies think of automation as some kind of panacea that will solve all of their production problems, only to discover that the automation has failed to live up to its promise. When automation fails, the follow-up assumption may be that the failure was the result of poor management or training. Both assumptions are wrong – the process was not prepared for automation. The path to world-class production is through the proper combination of the two technologies, and the order of each is of paramount importance.

Automating a process before simplifying it creates waste that will have to be eliminated later, at a much greater cost. Many producers prefer bigger and faster machines when they should be pulling the waste out of our processes first. When they have simplified their processes to the point that all waste is removed, then and only then, should they automate. They are only half right when they think that super technology is going to make them super producers. The gains that companies make through simplification and the removal of waste will be much larger than the gains achieved through automation. Managers have to understand that throwing money and resources at their problems will work no better than just **indiscriminately** cutting heads to meet Wall Street expectations. Both are short-term solutions that have long-term disaster associated with them.

Size does matter to many production managers. They want the biggest, fastest machine out there. Too often managers fail to understand that sometimes a couple of machines are much better than one super machine. When that stand-alone machine breaks, production comes to a halt. If they are told that they can count on 90% or better up time on a particular machine, they don't realize that the 10% downtime can be disastrous **if** it hits them at the wrong time.

This is a mistake producers make much too often. We almost insist on big, powerful stand-alone machines that we buy to give us greater capacity. When these stand-alones break down, we are in a pickle. If most or your product goes through that machine, there will certainly be a serious problem if that machine has a serious breakdown. In most cases, the proper approach is to purchase several smaller machines. The breakdown of a smaller machine will not be as catastrophic as the breakdown of a large stand-alone machine. There will be a loss of output, but it will not be a **complete** loss. The producer of that big machine may promise 95% up time, but think about it, do you think that the 5 percent down time will hit you when you can least afford it?

Machines are often purchased when they are not really needed. Before that super machine is purchased, make sure that the process really needs it. It may be possible to improve the output and performance of the machines already in place. More often than not, we buy those big hummers that pour out a million widgets an hour, when the rest of our processes can only handle a hundred widgets an hour. We find it irresistible to be the baddest of the bad, even when it comes to the machines we buy.

One of the most critical issues affecting long-term productivity is the argument over the allocation of space for new machines and work centers. More often than not, when the facilities department is given the sole responsibility for determining where to place new machines, or expand a work center, they look for available space, often ignoring the effect the new location will have on the effectiveness of production. They do this because they really do not understand the business of production. They might think they do, and in some cases they may have a lot of experience with production, but they are not down there in the trenches doing the work. They are not the ones that have to meet the production requirements. Their only responsibility is to find a "reasonable" place to put the equipment.

They are usually not educated in what makes a production process successful. They will often put a new machine in a place that will increase the amount of work that it takes to put product out the door. They do so innocently, but their innocence is no less destructive.

When expansion (growth) is not planned and conducted properly, with consideration given to process excellence, the ramifications are not always instant, but they are definitely predictable. Cycle time will go up **big time** because there is usually a profound increase in product transport required. Everyone has to work harder and WIP is increased because the location of the expansion is not right for manufacturing. Proper layout of the facility is not given the consideration that it should be given because the trump card is given to the issue of "available space" or short-term cost, not to process necessity and long term effect. The net effect of this type of reasoning is that the chosen location is a compromise, not an optimum solution. The next addition or expansion will add more waste and the downward spiral has begun. Productivity is not, and will not ever be a consideration because once this malaise begins; it builds on itself and gets to the point where it is virtually irreversible. Eventually, it will be totally out of control. Whenever growth is necessary, available space becomes the only criteria that will be considered.

If you were a growing company you would be very wise to prepare yourself for growth by establishing standards that will protect the productivity of your process during growth. If your process is already scattered to the four winds, and my advice is too late to help, you'd better give strong consideration to a complete layout change, or the creation of two identical lines. Do not start the process fragmentation that will eventually bring you to your productive knees. If you bite the bullet and accept the pain you've already caused yourself, you have an excellent opportunity to minimize your losses of output and revenue. But you have to understand that where you are headed with this fragmented process you've started is not where you want to be. I've seen this happen many times, and believe me; you do not want it to happen to you. Grow, but grow sensibly with respect for your present and future process needs.

Whether you bite that bullet or plan your growth, you will find that it is infinitely more rewarding than continuing to watch your process fall apart little by little. Once you have a smooth running process with very little waste, product transport and cycle time, you will recoup your losses and will most likely even improve product quality significantly. The most common argument against investing in a re-layout is that customer satisfaction will suffer because you will lose too much production time. While it is true that production time will be lost, it can be minimized if the re-layout is planned and executed properly. The second most common argument is that the cost will be excessive. It all depends upon whether momentary pain is preferable to a slow death. If it is, (and it should be) then get out there and straighten the process out.

When a process is compact and smooth, it almost always requires fewer resources and always produces more, or produces required levels more easily. In addition, those working there will be happier about their jobs because people enjoy working in an efficient process. Most importantly, your customers will be much happier because they will consistently get their product on time, with the quantities and quality they expect. The internal **and** external customers will get what they want, when they want it, with the quality they expect.

Although CFM shared many of the same values and tools that were used in JIT, it added some interesting new wrinkles of its own that helped to make it easier and quicker to implement their system. One of the CFM implementation tools that eventually became an important ingredient of Fast Cycle Production is TAKT. Takt is a German word that essentially means **beat**, as in the beat of a drum. Beat the drum slowly, and you have a slow beat. Increase the tempo of the beat, and the opposite is true. In production, you want the beat to be fast and continuous with outstanding quality.

If you think of beat as output, be it the output of the factory, or more importantly, a work center, the faster the beat, the more output you have. But, in order to have a rhythmic beat, with respect to output, you also have to be able to record the output periodically to make sure that you can measure your beat. A constant and predictable beat is important in a production process. Erratic output and product build-ups are destructive to production.

By posting your output each hour, you have an excellent idea of what your "beat" is, in terms of numbers, but that is not enough to make TAKT and TAKT charts valuable tools. You need something that tells you what is either holding you back, or Detractors, and what is working right, or Enhancers. By associating low output readings with a Detractor, and high output readings with an Enhancer, you can develop an understanding of just what you have to do to improve your output. The dynamics of TAKT Management are discussed later, but for now, it is important to know that what I call the greatest continuous improvement tool EVER is so much more than an output report card. The recording of hourly output numbers does not make it strong. Neither does it derive its strength through the recording of Detractors and Enhancers. It is the combination of those elements and the recording of that information in a database that is reviewed periodically that makes it so powerful.

During the implementation of CFM, TAKT charts were used to show how well things were going with respect to output and issues affecting output. It worked great because it added discipline to the movement of product on the line. If you are forced to ship all that you have produced for the last hour by the beginning of the hour, you are going to have a disciplined flow of product. A disciplined movement of product while working out the problems that restrict flow eliminates the need for large amounts of WIP. Why would you need a large amount of work lying around when you know that you will get all that you need for an hour at the beginning of the hour? It's very simple – you only need a lot of WIP because the delivery of work is unpredictable, or as was explained earlier, you have an internal delivery issue because of a fragmented process.

In a fragmented process, you have fewer deliveries because the deliveries are more expensive. One company that I worked with had spread its process out so badly over time that it was necessary for them to hire a good number of people that did nothing more than transport product from work center to work center. They had people that spent their entire work career just transporting product on mechanized carriers. Work centers that should have been placed side by side, were placed so far apart that it took 10 minutes to transport the product with a mechanized carrier. It was the worst situation that I have ever seen with respect to unnecessary transport needs. When the improvement team made a very strong case for layout changes that would have improved the situation dramatically, the facilities department effectively stymied the proposal with help from certain influential engineers. Every single person that reviewed the changes from the department managers to the plant manager supported the proposal and yet the changes were never made. The strong business case made no difference.

Even though we were unable to get the company to approve our layout plan, we were still able to accomplish hourly deliveries and implement TAKT Management. It was a monumental effort that took a great deal of planning, training and coordination by the Core Team with the departments and the transport people. The effort was successful, thanks to the hard work of the employees and the Core Team members, but it could have been accomplished much more easily with much less cost. I learned a lot about company politics with that experience. It showed me that presenting a strong business case is sometimes not enough to win the battle and that executive management is not always in control of its own processes. We were still able to establish TAKT Management in this plant, a credit to the workers and the team.

In TAKT Management, posting the information on charts, and recording the data in a database is not enough. Somebody has to look at the information and use it in a positive way. With TAKT Management, reports are disseminated to supervisors and managers. Daily, weekly and even monthly meetings are conducted to study the information in the reports and take positive action to eliminate the problems and exploit the opportunities that the reports expose. These meetings are a way to solve issues that just won't go away without intervention. Productivity is always improved when the obstacles to it are eliminated or minimized. When these meetings are taken seriously, and those attending work to find the cause of the detractors, amazing things can happen.

On the negative side, I have seen a tendency for some companies that have benefited greatly from TAKT meetings, to turn them into very unproductive sessions that look only at the "numbers" and ignore the reasons that the output numbers fluctuate. Some managers want to show the hourly amount that is needed to make their requirements on the charts. They don't seem to realize that TAKT meetings are forums for the achievement of continuous improvement, not a place to put pressure on employees to produce more.

On more than a few occasions I was called back to find out what went wrong with the meetings, only to discover that they no longer achieved their original intention. For some reason, these companies allowed TAKT meetings where the focus is on finding and fixing problems to be transformed into confrontational production meetings where the focus was on finding out who was making their output numbers and who was not. At some meetings Production Control people actually dictated which orders needed to be "prioritized". Management reverted back to pressure tactics because it was much easier than actually working to improve their processes. Why work to improve when you can apply pressure and exert power?

Management forgets what made them successful and regress to a policy of just pushing for higher numbers rather than working for them. Pushing for numbers is just a way for managers to exercise their authority. Using TAKT meetings to understand the causes and working to eliminate or minimize them requires more work, but it is much more rewarding. Pushing the numbers is always counter-productive because the focus is on people, not process. It assumes that people will produce more if they are pushed.

Over the years, I found that in every single case of a CFM implementation that I personally observed, the Takt Meetings changed very quickly after the implementation was complete, and it was not long before the meeting itself actually went away. The successes realized as a result of the use of Takt made it quite obvious to me that TAKT was the most effective continuous improvement tool that I had ever seen. I was so impressed with Takt as a continuous improvement tool that I knew even then that it would be a key element of Fast Cycle Production. The challenge I faced was to find a way to make it permanent, once implemented.

As a continuous improvement tool, TAKT has no peer. In each and every CFM implementation that I had been part of, the use of TAKT charts was instrumental to the success of CFM. TAKT Management has consistently produced outstanding results because it is more than a tool to be used during the implementation of FCP; it is designed to be permanent. The good news about TAKT Management is that it is the most powerful continuous improvement tool that a company can use. The bad news is that it is very fragile because it relies on management to maintain the purity and usefulness of the tool. Because of this dependence upon managerial mettle, the effectiveness of the tool could deteriorate over time because many managers just take the initial gains and then lose interest in the tool itself.

It is not that managers lose interest in continuous improvement; it is that they lose interest when their continued involvement is required. Some just want recognition for the initial gains and then want to go on to something else. They pass the responsibility for maintaining TAKT Management to people that are either not properly motivated, or do not have the authority to keep the system in place. Without the continued involvement of managers, especially at the senior level, there is a strong possibility that TAKT Management will eventually lose its usefulness.

With all of the outstanding tools contained within CFM, how is it possible that the gains would be temporary in each and every case? Why did CFM disappear in every single implementation at every plant where it was implemented? Many of my associates believed it to be associated with the IBM culture. That statement cannot be true because the same phenomenon has occurred in non-IBM plants. In the past, IBM managers moved from assignment to assignment with rapid regularity. Some say that the letters "IBM" stood for "**I**'ve **B**een **M**oved." When new managers come in, they usually want to "Do their own thing" and end up not giving priority to something that was created by their predecessors. Also, they might not understand the system that was put in place or believe in its usefulness. Regardless of the reason for CFM dying within a year or so of its implementation, the truth is that it did. If you ask those who were there when it came and went (and I did), you get no answer. They just do not know why it died. **I know why - It died because TAKT died!**

My discovery of the reason for the consistent demise of CFM would help to form what I call the **Continuous Improvement** element of Fast Cycle Production. In my opinion there are five major reasons for the consistent demise of Takt after a CFM implementation:

1. Top-level and middle managers either take a new assignment or they lose focus. Their replacements don't understand it or appreciate it.
2. Management never really believed in Takt as a continuous improvement tool – they merely tolerated it.

3. TAKT was always considered as an implementation tool – it was never thought of as being permanent even by the implementers.
4. TAKT meetings are either discontinued or are transformed into production "push" meetings.
5. All of the above.

It is quite clear that three out of four reasons for the premature demise of CFM are directly related to TAKT, and all four are directly related to management. TAKT was either eliminated, or it became something that it was never meant to be. The managers who supported TAKT initially either went away or were replaced by managers who could care less about it, or the original managers abandoned their support of it. When TAKT is maintained as a way to solve problems, everything goes well. When it is sustained, everything goes well and the company continues to improve. Any rational person could make a strong argument that reason number one causes reasons two and three. I am a rational person. Therefore, reason number one is the big kahuna.

Right now, some of you may have come to the conclusion that I am anti-management, or that I am overly critical of managers. I don't believe this to be the case. As I said before, top managers do not acquire their positions by being incompetent dolts. They obviously bring more than political savvy to the table. My intention has always been to convince managers to become advocates for process excellence and to maintain the pressure necessary to keep continuous improvement healthy in their plants. Of course, it is not the job of executive managers to be involved in day-to-day process issues. They have so much more to do than get involved in that level of detail. Besides, they have other people to do that. But what they can do is to maintain strong executive pressure to maintain the purity of the philosophy and tools that are in place. It won't take much of their time, and the rewards will be profound. If they insist that focus be maintained by their reports and periodically follow up on the progress, I'm sure that the focus will be maintained

The reason I decided on the name: **TAKT Management** (TM) was that the use of this tool should be a way of life – a constant presence - always there and always working. In other words, you should use it to **manage** the process. To effectively utilize TM, you must have a way of learning from the past. The only way you can learn from the past is to be able to record data and then go back and look at it. The best way to archive and recover data is with a database program. A good database program allows for the easy recording and retrieval of data in a useful way. Customized reports can be created that allow you to analyze data in a meaningful way.

My first use of a database with Takt was a huge success! It just so happens that this particular project was also my first application of TAKT Management. On this project we continued TM beyond the implementation period and actually created a database program during the project. Once we got enough data into the database, we printed and distributed reports to supervisors showing the impact of certain Detractors and Enhancers. We were able to create reports on production that not only showed the negative effects of detractors but also the positive effects of enhancers. We could also show which detractors were showing up most often. Each line area could now quickly learn its "Top 3 Detractors and Enhancers" AND quantify the effects they had on the line. That told us which element we should work on and the order in which we should work on it. We really got creative and the results were fantastic. I had found another element of Fast Cycle Production! As it turns out, TAKT Management is **THE** major tool in FCP.

Once the distribution of the reports began, we kicked off our weekly TAKT meetings. At the meetings, we assigned the responsibility of working on the top Detractors and Enhancers to supervisors. The result of this was that some really outstanding "fixes" started taking place. One after another, problems were solved and roadblocks eliminated. The weekly meetings were a huge success. Problems were being solved. However, we found that getting to the source of some of the problems was pretty difficult.

We found out that there were still some problems that lingered around too long. They were like bad pennies; they just wouldn't go away, despite everyone's best efforts. It was time to involve upper management in the process improvement business, so the monthly meeting was born. Upper management wanted to get involved on this particular project and was excited about what was happening. What they did not want to do was sit there for hours and just listen to presentations so we had the middle managers make brief presentations of "aging Detractors". They also were unable to attend most of the daily and weekly meetings so we created the monthly TAKT meeting for executive management. Appropriately, upper management used its influence to solve some of the problems. Once the word got around that upper management was an active participant in process improvement, even more problems got solved.

In some cases, it was difficult if not impossible to get upper management to commit to monthly meetings, due to travel and their involvement in other meetings. Because of the inability of senior managers to commit to regularly scheduled monthly meetings, weekly meetings became the only meeting other than the daily meetings. It is preferable that supervisors are the dominant contributor at weekly meetings, but in some cases, middle and senior managers took control. Over time, in each case where we were unable to get upper management commitment to monthly meetings, there was some deterioration of the effectiveness of weekly meetings. The system really works when executive management is involved in TAKT Management.

Even though the results of TAKT Management were more than impressive, it was not long before I discovered how vulnerable it was to recidivistic behavior. All it took was one middle manager working to turn the meetings into the production meetings of old, where people were pushed to produce rather than solve problems. It seemed as though certain members of management were never really comfortable with continuous improvement because they themselves could not take credit for it. They could take credit for pushing for more output. As a result of their misguided efforts, there was a return to power management.

In a return visit to a plant that had implemented FCP with outstanding results, I discovered that the daily TAKT meeting no longer even resembled its old self. The meeting was not even a good imitation of what it used to be. The charts were outdated and incomplete and were totally ignored. Nobody paid attention to the detractors and enhancers; they only paid attention to the output numbers. You could have put an obscene statement on the chart in place of a detractor and it would have gone unnoticed. They had regressed so far that they almost looked like they did when the initial project began. They still called their meeting a TAKT meeting, but the resemblance was purely coincidental. It was sad for me to see such serious deterioration of TAKT meeting effectiveness.

Recidivism is quite often associated with criminal behavior. People who have led the life of a criminal are rehabilitated, or so it is thought. For a while they are leading a good life and doing the right thing. Then, for absolutely **no logical reason** they revert to their previous behavior. They know, of course, that their behavior is destructive and that they are heading in the wrong direction but they revert anyway. This is what happened at the plant of which I spoke. They had regressed to a previous pattern of behavior that was not productive even though they had been successful and productive prior to that. Here was the CFM phenomenon back to haunt me. In this case it took two years instead of the usual one-year for CFM to disappear, but disappear it did. I wondered why it was always so difficult to sustain success. Was there something that I was missing?

Guarding against recidivism was an ongoing battle for me. I went from project to project trying to take every precaution I could to keep it from happening. I knew what type of person was most likely to be the catalyst for recidivism. Technicians, leads and supervisors were very likely to want to change the meetings; after all, they were the ones who were hit with the action items. There was also resistance from middle and upper level managers who either didn't believe in TM or didn't really understand it. Furthermore, there was a danger that sales and production control together would turn the meetings into a forum for promoting their agenda: prioritizing orders to satisfy their concept of what customer service is all about.

Whenever a successful productivity improvement project takes place, dramatic improvements follow soon thereafter. Everyone is happy, and upper management gets accustomed to seeing huge production gains take place. High fives, promotions and rewards are the order of the day.

This positive attention is habit forming. People have a strong desire to be seen as successful, and managers are no exception. It's easy to look at the improvements and be prone to make predictions such as, " Based on our productivity increases, we will be able to increase our output by 1000 widgets weekly, resulting in an increase in revenue of ... dollars." The promises may come easy after a successful implementation, but it is quite often forgotten that the promises can only come true if improvement continues. People have a hard time understanding one important thing about manufacturing: Sustaining process improvement is more difficult than achieving it. It is amazing how easy it is to lose all that you have gained through nothing more than neglect.

In a growing business, where you can sell it if you can make it, an increase in productivity is naturally translated into a commitment for more output. When manufacturing starts bragging that they have cut their cycle time and increased their output and productivity, the drooling starts. Once managers say that they can make more, it is a natural to assume that they will have to belly up to the bar and back up their promise. Salespeople make commitments that they expect manufacturing to meet. Soon the extra orders start coming in and all eyes are on manufacturing. As long as manufacturing rightfully follows established run rules and continues to maintain their values, they will be able to meet the increased demand. All that they have to do is to continue doing what they have been doing! If they panic and go back to old practices, such as prioritizing orders, they undermine their abilities and pretty soon they are prioritizing everything.

This downward spiral need never happen, but it does when values break down. The next thing you know, a promise is broken. In order to "keep the customer happy", some manager starts a "Hot List" which tells the line people what to work on and in what order. Pretty soon there are hot lists all over the place. Forget about TAKT and improving your abilities – you're just trying to survive here. Suddenly people start wondering how this could have happened to them and why they were doing so well and suddenly everything is falling apart. It must be the fault of (take your pick): JIT, CFM, FCP or TAKT Management). Whatever system they are using must not be all that it is cut out to be. Blame the system when you should be blaming the person you see in the mirror.

To pacify the increasing amount of concern by the customers for getting their orders on time, management will start increasing the amount of work that is released to the floor. Boy, there's a bright move! If you build it, they will come. No, I mean if management releases it, it will be shipped. That's true of course if you are talking about next month. By releasing too much product to the floor, management has just reduced the amount that the floor will be able to ship. They know (I think) that if they release more product than the process can produce, that it is just going to sit there, but they release it anyway. Any time that I go into a factory and I see the presence of hot lists, I know that there is too much work on the floor and the situation is most likely deteriorating, not getting better.

Prioritizing orders is a downward spiral that just keeps getting worse and worse. Prioritizing can be the end of TAKT Management and its demise can come quickly. Prioritizing product has never worked effectively in the past, is not working effectively anywhere right now, and never will work in the future. I cannot say it more emphatically than that. A certain amount of prioritizing can be tolerated (which I will cover later), but the ability of the process to move product efficiently makes that determination, not some member of management.

Healthy Culture – Healthy Company

While I was working with a client in Colorado, I met Gerry Clarke, a consultant working with the same client. He was doing his thing and I was doing mine. Clarke and I eventually became partners in a consulting company. We would work together for about five years as partners before Clarke would become the president of the company whose cultural enhancement product he had been using. Clarke had formed a relationship with, and was a qualified consultant for Human Synergistics. HS developed a set of tools to effectively determine the culture that exists in a company. Clarke was teaching his clients how to identify and correct destructive behavior and improve their cultural makeup.

Clarke used the inventory tools developed by Human Synergistics to measure the health of the culture, to identify "hot spots" of destructive behavioral patterns and to identify an improvement path. Once the people associated with the cultural abnormalities were identified, counseling could take place to replace the destructive behavioral patterns with healthy ones. Before we met, the cultural changes that moved a company from destructive to healthy behavioral patterns took years to accomplish. That was until Clarke and I worked on our first project together.

The first project that we worked on together gave us the opportunity to do something that I had wanted to do for some time: combine cultural development with a process improvement effort. We called it, "The Combined Offering". It was a seamless application of both of our technologies and we applied it at a plant that was in serious trouble in both areas. When we first started that project, we knew that the plant was in trouble on all counts, but we did not know just how bad their culture was.

It probably was not the worst culture that Clarke had ever seen, but it was serious. In spite of the challenge, we had guaranteed the company that we would make their process better and reduce their cycle time within six months or we would forfeit some serious money. Now let me tell you, trying to implement FCP in a plant as culturally bad as this one was a serious challenge. In order for me to get things done, I would need the cooperation of a lot of people. Clarke's job was to transform that culture into a much healthier one.

The plant was doing very badly on a number of fronts, in addition to the serious production problems that they were having. Furthermore, they had the worst factory layout that I had ever seen. If these problems were not enough, we also had to contend with the following:

1. Morale was miserable, to say the least.
2. Most employees were working 7-day workweeks and still not meeting expectations.
3. Production meetings strongly resembled a vicious kangaroo court. They lasted a long time – time that could have been spent producing product. Employees were often belittled for their inability to meet production numbers. They dreaded going to the meetings.
4. Management was more likely than not to work 80-hour weeks. Meetings were held late at night and quite often turned into screaming sessions.
5. Creativity in the work place was nonexistent. Workers were too scared to do anything but try to stay out of the spotlight.

I could go on, but suffice it to say that the plant was a horrible place to work. We were there to help these people improve their ability to produce product. Although some of them really wanted help, most of them saw us as nothing more than a temporary distraction that would just make it even more difficult for them later.

Clarke and his people went to work on the underlying reasons for the cultural breakdown. They worked to improve the employees' behavioral patterns and helped them to understand what they could do about their destructive behavior.

The combination worked dramatically. I taught them the art and science of manufacturing and Clarke taught them to accept change and to develop a healthy, properly motivated culture that thinks in humanistic encouraging, self-actualizing and achievement oriented ways. His accomplishments made my work possible. From that time forward I have believed in always including cultural development in a productivity project.

There were some discoveries for Clarke also. Up until this time, experts had always maintained that significant cultural change took five years or more to achieve. It was considered impossible to accomplish a cultural turnaround in a year or less. Well, we found out that the experts were correct to a point – you can't do it without a catalyst.

The catalyst in this case was called success. When people who have been failing suddenly find that they **can** be successful and that their effort does make a difference – a positive difference - they find themselves solving problems. The Combined Offering accomplished more than I ever imagined. I remain amazed to this day that I was ever able to accomplish what I had without this wonderful tool. The combination of productivity and cultural enhancement was a marriage made in heaven.

Looking back with 20/20 hindsight, I can see that our discovery was as significant as **anything** that I have learned to that point. The combined offering was a lot like positively treating the body **and** the soul of a person at the same time. I truly believe that it was the combination of the two that made this plant go from being on the brink of destruction to being profitable and successful. If you could have seen this transformation you would certainly agree that it was truly magical. I have never seen a turnaround as dramatic as this one.

As incredible as the tools were that Clarke used, they were not more impressive than the man himself. Clarke helped everyone that he worked with, including me. I have always been very inflexible when it comes to my work because my work required that I be that type of person. I have to provide leadership. Clarke taught me how to do that in a more positive way. I learned that there are better ways of getting people to accept what I was trying to teach them.

To that date, I had concentrated on building a process improvement methodology that could be implemented successfully and sustained indefinitely. I learned from the pioneers of modern manufacturing and adopted many of their values and tools. Although I modified many, if not all, of the tools to make them more effective in modern processes, I did not modify any of the pioneers' VALUES because they are as appropriate today as when I first learned them.

I have learned what makes production work, and I know what kills it. I also know that if you do not manage your process in a way that convinces your people that you are committed to their success, you may get work out of them, but you won't get their dedication to **your** success.

Managers too often work only towards their own success. They grasp every accomplishment and present it to their management to show how successful they are. When things are not going too well, they often either find excuses, (usually they blame someone else) or they keep the mistakes hidden. Too often, detractors are not something from which to learn, they are things for which to find blame.

A very wise person once explained two types of simplicity that can be used to arrive at and implement a solution to me: Right hand and left hand simple. Left hand simple solutions are the types that have very little value because they are a relatively mindless form of simple. It is like not putting your hand on the stove as a solution to not getting burned. Another example of a left hand simple solution is to make it a policy of counting twice when there is a counting error. It might cut down on the number of counts that go wrong, but it is a solution that has more cost than benefit. A right hand solution, on the other hand, is one that would seek a solution that fits within a proper value set. This is a solution borne of thought. The objective is to create less work, not more, so the trick is to find a solution that is based on eliminating the root cause reason for the bad counts in the first place. Now you're working towards a right hand simple solution.

In the Solutions chart, as you move to the right on the x-axis (horizontally), the complexity of the solutions increases with a corresponding increase in cost. The ratio of complexity and cost are not really linear, but for the sake of illustration, they are shown as such. Suffice it to say that, as the complexity increases, the cost of the solution goes up. Complex solutions require more cost because they typically involve equipment purchases, engineering costs, resources, time, money, etc. Complex solutions also take a long time to get implemented. With a complex solution, the cost may be measured in dollars and the time that it takes to solve the problem is extended.

Moving from left to right on the x-axis, the solutions become more complex until you reach the pinnacle of complexity (and cost). Then something wonderful happens – the solutions become simpler and simpler and cost less and less. But, this is a different kind of simple. This is a simplicity borne of thought. This is a simplicity that produces workable, effective solutions at less cost. The brilliance of the solution is in its simplicity, and the further you go to the right, the more simple and brilliant the solution is. I wouldn't give you a penny for a left hand simple solution. They end up hurting more than helping. But a right hand solution is priceless. With right hand solutions you get results that last. Simplicity born of thought – what a concept!

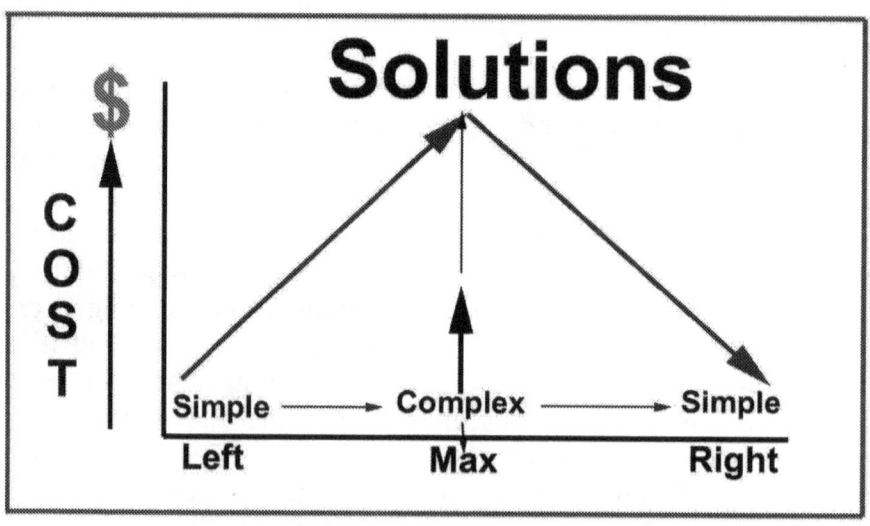

Figure 1.7.1 – Right and Left Hand Simple

Figure 1.7.1 illustrates the concept of Right and Left hand simple solutions. Starting on the left, the solutions are inexpensive and simple, but they are a very meaningless. They are cheap to implement, but could cost a lot of money later. As you move towards the right, the solutions become more complex, meaning they require study and coordination and are often expensive (capital purchases). Eventually, you reach the height of complexity and cost. Examples of these are major purchases of expensive machines.

A very good example of a right hand simple solution is Kanban WIP control. I will explain the uses and effectiveness of Kanbans, and a creation of my own – Dynabans - in a later chapter, but for now it is important to know that Kanbans were created to solve a very serious problem in production: excess WIP and the problems that are caused by excess WIP. Kanbans are effective line tools that control the amount of WIP on the line; they are very easy and inexpensive to implement, and they really work! They are right-hand simple. Another right hand solution is TAKT. It was designed to establish disciplined movement of product while providing a way of identifying and solving problems on the line. It is easy to implement, costs very little to implement and produces big results. The beauty of the solution is in its simplicity – A simplicity borne of thought.

I have seen some left hand simple solutions in production that were initiated by some very well meaning but thoroughly misguided people. Some of these so-called solutions have stuck around for years! There are companies where these mindless procedures have been in place for so long that nobody remembers who came up with them or why. Most of these so-called solutions are the result of promulgation. A left hand simple solution is most likely a punishment to someone that goofed up and did something wrong. Somebody makes a mistake and right away some supervisor is requiring a double check from then on to make sure that the "stupid" mistake is never repeated. The only thing stupid here is the solution to the problem, not the mistake.

I picked on double counting as a solution to a missed count on purpose. This is one of the wasteful process steps that I have seen almost everywhere. With this solution, logic tells you that three counts are even better than two. If you really want to do it right, why not make it four? It is pretty obvious that the genius that came up with that rule really put a lot of intellectual effort into it before making a decision. Even though I have made fun of the solution and the person that came up with it, my jest is not intended as a personal attack, but is intended to emphasize that solutions of this nature are counter-productive at best, and destructive at worst.

If you think I'm being tough on this issue and on those of you that have come up with solutions like this, you're right. I have a very important reason for doing so: If you come to believe in your heart of hearts that redundant counts or inspections or redundant anything is truly wrong and destructive to your process, you will continue to look for a solution that really works. You will see them for what they are: waste. You will want to eliminate waste, not perpetuate it. If I have convinced you to start looking for proper and meaningful solutions to problems, I will have accomplished my goal. By showing the insanity of forced redundancy I may have prevented the implementation of these types of punishment-oriented solutions.

Anyone who comes up with these kinds of solutions is targeting people as the culprit, not looking for the true problem and working towards a meaningful solution. Consider the rabid thoughts of a supervisor that has been chewed out for some miscounts: "If they can't count right, I'll fix their wagon – I'll have them counting everything twice. Then to make sure that I catch them when they goof up, I'll make them record their counts on this official chart so I'll have evidence that they did it wrong. Better still, I'll require them to stop somebody else from doing what they are doing and come over and do another count. I'll make it sound humanistic by calling it 'The Buddy System'. If I do all of this, I will be successful because my department will never get hit with a bad count again."

How can we improve on this unfortunate set of procedures? First of all, we have to start out with some new assumptions and we have to believe in them:

1. People are not the problem. They want to do a good job. Get a grip and understand that! If you disagreed with that last set of statements, then I can tell you that you are right – a person is the problem and that person is you.

2. The process **is** the source of the problem. If somebody miscounted, we should look for an easier way to count. (The solution here does NOT include giving math tutoring to employees.)

3. We are not in the business of counting; we are in the business of producing product. Is the customer willing to pay us more for the product if we count more? If we're not in that business, then we definitely do not want to increase the number of counts.

4. If the count is so important, and it may well be if you are dealing with an expensive product, make it impossible to

miscount. I'm not talking about hiring math majors as employees I'm saying that you should find a better way of counting that makes it impossible or highly improbable to miscount.

5. When you are working to eliminate redundant counting, do not just do away with it. Eliminate the reason for redundant counts by replacing it with a right-hand simple solution.

6. Did I mention that people are not the problem?

In short, find a solution to a problem that is people oriented and does not cause the employees more non-value add work. Involve the people in the decision process instead of going on a power play. If you are really interested in the success of your company, you understand that your people need managers that work to make them successful. Managers can't make employees more successful by adding wasteful tasks to their work. If you are a good manager, you will try to make your employees' work simpler, not more complicated. If your mind goes directly to "correcting the workers for their incompetence", your values need adjusting, just as mine needed adjusting at one time.

Managers – You can't live with them ...

I've been told since the day that I taught my first class on Fast Cycle Production that I was always very critical of management. Actually, managers have told me this. I often talked about how management created the problems that the line workers had to live with. I would say things like, "Management shackles our arms and legs and then tells us to run a four minute mile." I've made other endearing statements to managers and found that most of them take the criticism pretty well. There are some who are sensitive to my criticism; most of the time they are managers who have promulgated wasteful procedures. It might seem strange to some, but I'm really trying to form an alliance with them, not trying to alienate them. I just have a funny way of showing it.

When I am critical of senior managers, I do so because I want them to understand the profound effect that they have on their employees. They have this effect because they manage the processes where their people work. I tell people that it is absolutely necessary for management to be united behind one manufacturing philosophy. When management truly believes in and supports one philosophy, those whose work lives they have influence over will support it also. I want to be able to convince management that fragmented values are nothing but a series of compromises that always makes things worse for their people.

I know this because I am no different than most people out there. Those that I admire and respect have the ability to influence me and always have. Like everyone else, I watched my manager for signs that I was doing a good job. I not only listen to advice from certain people, I solicit it and am inclined to accept it. Most of us understand that the people that manage us are fallible and subject to occasional mistakes, but we still want to be seen by them as capable, or even better, outstanding at what we do. Coming up with procedures that are insulting and do nothing more than create unnecessary work is not the right thing to do. Workers need to be guided by a better set of values.

One of the best lessons that I have learned since I began consulting was that managers at all levels quite often fail to realize that everything they do is under scrutiny. Every word they speak is not only heard by others, but it is also translated and passed on to others in the newly translated form. Not only that, a manager usually does not even have to say a word … all they have to do is to frown, smile, laugh, shrug their shoulders or look up in the air and the action is observed, translated, and passed on to others.

Many managers are trained to choose their words carefully because they know that someone may translate their words or actions in the wrong way. If managers allow their thoughts or emotions to betray them to others, they are thought to be weak. Managers know very well that a certain reaction in a meeting or on the floor could cause actions to take place that may not necessarily be in the best interests of the company or may even be destructive to their process. It does not mean that they should be phony and hide their approval or even disapproval when it is appropriate to do so. It only means that there are many occasions when it just may be counterproductive to react openly when momentary silence would give a manager time to evaluate a situation. If, on the other hand, managers consciously and intentionally send out signals or hints rather than clear instructions, they are probably being manipulative. Manipulative behavior is destructive.

When there is a great deal of prioritizing taking place on the line, an employee could push an order ahead of others based on an innocent question or statement about that order from a supervisor or manager. The person hearing the statement could decide that the manager is subtly asking him to push that particular order ahead of others. In an executive training session that I conducted, a high level manager confessed to the rest of the class of an occasion when he was talking with one of his supervisors about production issues and innocently asked the supervisor about an order that had experienced delays. He told the supervisor that he had received a call from the customer asking about the order. All he wanted to know from the supervisor was whether the order was moving normally. He was not trying to get it pushed ahead of other orders.

The supervisor did not hear the question as an innocent one, he heard it as an order to get that order moving at all cost. Since it was not presently at the front of the line, the supervisor went to his people and told them to give the order the highest priority. Since the people on the floor wanted to support their supervisor, they stopped everything they were doing and pushed all of the work in progress aside to work on the order. They were happy to report that they had been successful and the order had been shipped in record time. When the manager heard what had happened, he was astounded that a very innocent question had produced such a response. This could have been a happy story, but it was not that at all.

The supervisor's reaction to the manager had been costly. The time it took to tear down the line, put all the WIP aside and start working on the new order caused every other order to be delayed. The teardown had caused even more problems because some of the product was partially finished. There were quality problems with work that was sitting around. The work that was put aside took longer to complete in its half-finished form than it was normal for new product to be completed.

Surprisingly, there were even run rules in place that should have stopped the teardown. The run rules stated that teardowns were strictly prohibited. Unfortunately, since it was the supervisor who overrode the run rules and directed the action, the workers did as they were told. The supervisor knew about the run rules and disregarded them because he believed that his manager was ordering the action. Both of them were surprised when the repercussions came, but their surprise did not equal the confusion that the workers felt. If the supervisor could order the breaking of run rules so easily, were the run rules really valid? From that moment on, the run rules meant nothing to the employees.

I learned early on that prioritizing was a very costly practice. It is a major killer of production **in all but world-class companies** for a very good reason. But there are tools that can measure the "ability" of a process to prioritize without adverse effects. These tools are used to calculate the cycle time and the ratio of cycle time to raw process time. They are also used to calculate the appropriate amount of WIP for the line. The tools are X Factor and Little's Law. I will explain their use in detail later in the book. For now, understand that prioritizing one order will **always** have an adverse effect on the cycle time of other orders. To be otherwise is impossible. Prioritizing always has a cost, but through the proper management of the ratio of WIP to output, it does not have to be excessively costly.

I have seen companies use prioritizing as a way of life even when the prioritizing was clearly destructive to their ability to meet customer requirements. They stuck to their philosophy as if it was the only thing that they knew how to do. If prioritizing could be marketed, these people could dominate the market. They were that good at it.

Unfortunately for them, they did not position themselves properly by maintaining the proper ratio of WIP and output to pull it off. A few of these companies even sold what they called "Quick Turns". They promised their customers that they would deliver product at an accelerated rate. They charged more for this service and customers were willing to pay for the service. The results were not good. Whenever a problem hit, and problems always happened, it was always disastrous and costly. They lost money and never seemed to figure out why this was happening to them. They considered it all bad luck.

Who Taught Whom?

Going back to simulation modeling, I said that teaching simulation modeling taught **me** more than I was able to teach my students. I taught people how to take data, how to enter that data into a model and I even taught them what values to apply when looking for solutions. From them I learned that simulation could benefit any kind of company. The same is true with FCP. When I was looking for a name for my philosophy, I eventually settled on Fast Cycle **Production** instead of Fast Cycle **Manufacturing** because FCP is not just for manufacturing. It can be used effectively in virtually any industry. It was one of my students that suggested this to me, not a brainchild of my own.

The education that I received with my simulation business taught me volumes about the science of manufacturing but it did much more for me. My belief in the values of JIT and CFM were strengthened through my simulation experiences because I could see them working in project after project. These were time tested values that I knew were valid because I had seen them work first hand. I didn't just believe in all of that stuff, I had lived it, and the beliefs became ingrained in my value system. They didn't just work; they worked amazingly well every single time! It is no wonder that I am so inflexible when it comes to production values.

I learned about different kinds of processes and their problems out of necessity. I eventually became a walking simulation program. I developed the ability to walk into any type of company and be confident that I could determine what was wrong with their process and present them with a plan that would improve their productivity. It was through these abilities that I evolved from being a simulation modeling consultant to a productivity consultant that included simulation modeling in his methodology. How easy it is to fix something when you have the right tools!

I also learned something about overtime that should have sunk in while I was a production manager but didn't. It finally sunk in after I had worked with a few companies that were struggling to get their product out on time. In every company that is more likely to look at their problems and associate them with people issues, it is predictable how they will respond to production inequities. They will either look for more people, or they will try to get more work out of the people that they have. It is so easy for managers to take the "people approach" to problem solving.

Whenever a company suddenly finds that it is unable to meet production numbers, one of the first things tried by management is overtime. Overtime is easy to prescribe because it involves people. The logic is simple: If you're unable to meet your numbers, don't hire new people, just use overtime to make up the deficit. Although many companies have a policy that restricts the amount of overtime that can be used, the bottom line is that if the orders are there and they can't be shipped, overtime is the answer. It is such an easy solution to the problem – just put your people on overtime until the problem is solved and then end the overtime. The increased revenue that it brings in will cover the overtime cost.

The problem with the use of overtime is that it quite often does not turn out to be a temporary solution. It starts out fine and the overtime does produce results. If there is a successful attempt to improve the process capability, the temporary overtime solution is a good one. Unfortunately, most of the time management will just keep on ordering overtime. Soon, overtime becomes mandatory and people are required to come put in extra hours. There are attempts to "manage" this mandatory overtime, but the situation remains the same – people are working long hours.

After the overtime is sustained for several months (the amount of time varies from company to company) a strange phenomenon occurs. Suddenly, or not so suddenly, depending upon how you look at it, it takes 7 days to produce what used to be produced in 5 good days with occasional overtime.

I have heard some theories from managers and employees explaining the phenomenon. Some believe that employees become accustomed to the overtime pay, which essentially means that they are making more money and eventually match their spending to their inflated earnings. The employees protect that inflated income by producing less on overtime.

Another theory explains the phenomenon quite differently. Some theorize that the workers are opposed to the overtime because it gives them less time with their families and recreation. The loss of productivity is a passive response to the mandatory overtime. People get burned out from working so much and their productivity suffers as a result. Still another theory is that working efficiently is less necessary in a factory that has experienced extended overtime. There is no reason to be efficient because it won't spare you from working overtime.

All of these theories have possibilities, but I really believe that the true answer is really a mixing of all of them. As far as I know, there has never been a credible study that explains the loss of productivity in an extended overtime environment, and if there were, it would probably raise more questions than it answered. There may even be some disagreement that the phenomenon exists. My experience has shown it to be true in each and every case, including my own plant.

I believe that mandatory overtime will not yield significant results when used as a long-term solution. Improving the process is the only sure way of increasing productivity and improving output numbers. Sustained overtime is costly and the results are only temporary. Those that have found themselves in a long-term overtime situation know that what I say is true. The achievement of process excellence is the only meaningful way to meet production requirements, and to continue to meet them

The gifts that I developed over time, and I really see them as gifts, have enabled me to lead my students towards solutions that they themselves were unable to see. Maybe it is because my students are too close to their problems and have been living with them so long. They can't see the forest for the trees, so to speak. The problems they have been living with may have either become acceptable or invisible to them. I have seen opportunities on the line that I thought were so obvious that I found it incomprehensible that only I could see them. Was it because my clients had never known another way or because they were comfortable with their situation? On one project the client told me that they had attempted to put a team together to eliminate waste from their process and had made some progress but could not find any new opportunities. To me, the opportunities were like neon signs. How could anyone miss them?

I found that by collecting data and building models each of us on the team eventually learned more about the process being studied than the people actually working or managing it, and it didn't matter how much we knew about the process when the project started. Not only that, they also came to understood the effect that the changes they were testing would have on upstream and downstream processes. I learned that it is impossible to fix a process until the details of that process are understood. The details of the process cannot be understood until the process is mapped to the lowest level of detail. It only took one project to create a team of experts on the most intricate details of the process under study.

To really understand the details of a process, or as a prerequisite to building a model, you must first flowchart the process down to the lowest detail. Here is another area of training that eventually produces process experts because all of the details contained within the baseline model are established during the process mapping effort. In addition to flowcharting, data must be collected to determine the amount of time that the product is delayed at each process step, and under what conditions it will be delayed. Process mapping, like simulation modeling are not just tools, they are teachers! When used in tandem, their power is almost limitless.

Much of the FCP methodology was established while I was teaching simulation modeling and consequently, simulation became an important ingredient of the Fast Cycle Production methodology. Simulation enabled me to lead my clients to meaningful solutions even though I knew before the simulation was tested that the solutions would work. Time after time I looked at new companies and processes and saw the same problems. The company may be new but their problems were the same as their predecessors. The most rewarding part of my consulting life has been my ability to teach my clients how to recognize opportunities and associate those opportunities with appropriate solutions.

The Facts - Just The Facts

I define the *product* of a process as the entity that is eventually delivered to the customer. In the loan department of a bank, the product is the completed loan. In an automobile factory, it is the automobile. On a smaller scale, the product of a work center could be the steering wheel that goes into a car or it could be the creation of a clock in the same automobile. Each step in the process either contributes directly or indirectly to the completion of the product.

I define a *process* as a series of related discrete events working in a serial and/or parallel fashion with each discrete event consuming time and each dependent in some way on each other. A flowchart allows one to understand the flow of the product through the process. By adding certain details and dependencies such as decision points and alternative paths, you can fine-tune the flowchart. The flow of the product is only the beginning – the little details that govern the flow are essential to the completion of the flowchart. A completed flowchart with all of the details that enable it to be analyzed is a mapped process. The whole process is known as process mapping.

Analyzing a mapped process is pretty straightforward. Some process steps that are promulgated (added to the process through decree) that contribute very little or not at all to product completion are known as waste, or Muda if you choose to think in JIT terms. If the workers are required or feel the need to complete these undocumented process steps, they are included in the process. What I mean when I say, "feel the need" is that sometimes the workers have their own ways of doing things that are not known by management and are not documented. Workers may feel that the procedures do not make sense or they might just have been trained to do the job differently than the procedures show.

The same rules of process mapping apply to the manufacturer of widgets and/or the producer of loans. The rules are the same for mapping an engineering change order process, or the process of the test department of a printed circuit board line. These are all examples of producers of product, albeit completely different types of products. As long as **something** is produced which can be defined as a product, it will benefit from process mapping. I have worked in virtually every industry and have found this principle to be true. Any discrete event process will benefit from process mapping and simulation. It is for these reasons that process mapping and analysis is such an important part of the Fast Cycle Production methodology. I wouldn't go anywhere without it.

The remaining sections of this book will explain the details and applications of the values and tools contained within FCP. I will also explain how to implement FCP in a production plant and what the correct order of each step of the implementation should be. As I've already stated, I've learned a lot about how best to implement my system and I believe wholeheartedly that this system cannot fail if implemented properly. If implemented correctly, and rigorously sustained, it will produce results that will surprise and amaze you.

There you have it. I've given you the history and origin of Fast Cycle Production and how I personally came to believe in the values that I've adopted and strongly believe in. My methodology didn't develop overnight and each detail of the methodology is truly time tested. That's why it never fails. The ingredients of FCP worked for the masters that pioneered JIT and CFM and they worked for me. They will work for any company and any process within that company. Even though FCP began as a manufacturing philosophy, it evolved into a philosophy that could work for the non-manufacturing elements of manufacturing.

Now let's get on with how to make this methodology work for you. The following sections outline the details and importance of the values and tools included in the FCP package, and how to implement each of them. The book will not make the reader an expert in the implementation of FCP, but it will provide you with a very good understanding of the elements of FCP.

Part 2 – Modern Manufacturing Values And Tools
- Chapter 1 – Modern Production Values
- Chapter 2 – Modern Production Tools
- Chapter 3 – Process Mapping and Analysis

Part 3 – TAKT Management – The Best Continuous Improvement Tool Ever!
- Chapter 1 – Elements of TAKT Management
- Chapter 2 – Implementing TAKT Management

Part 4 – Modern Constriction Theory – Gate Busting
- Chapter 1 – Basic Constriction Theory
- Chapter 2 – Gate Busting – How to Deal With Constrictions

Part 5 – Setup Reduction – The Team Method of Setup Reduction
- Chapter 1 – The JIT Way
- Chapter 2 – The FCP Way
- Chapter 3 – Implementing Setup Reduction

Part 6 – Kanban and Dynaban Management – A Right Hand Simple Solution That Still Works
- Chapter 1 – Basic Kanban Theory
- Chapter 2 – Kanban and Dynaban Applications
- Chapter 3 – Implementing Kanban/Dynaban Management

Part 7 – Run Rules
- Chapter 1 – Run Rules Help To Maintain Process Integrity

Part 8 – Organizational Cultural Improvement
- Chapter 1 – The Birth Of The Combined Offering
- Chapter 2 – The Human Synergistic Tools

Part 2 – Modern Production Values And Tools

Tom Clason

Chapter 1 – Modern Production Values

First of all, let's define the word "values". The dictionary defines values as "A principle, standard or quality considered worthwhile or desirable." Another dictionary defines values as "a guiding principle or ideal." The definition of values that we will use is "Deep seated beliefs that guide us in our decision making process." Defining the term, "deep seated beliefs" is a little more difficult because we all have different opinions as to what constitutes a "true, or deep seated belief".

An individual can say that they believe in God, aliens, reincarnation or whatever and, in some cases, that belief can well be very shallow and insincere. In other cases, the belief may be extremely profound and unshakeable. People may say that they believe that stealing and cheating are wrong and yet they will cheat on their income taxes. To some, values are conditional. To others they are truly values – solid and uncompromising. Of course, when values are compromised there is rationale that is designed to explain it, but the rationale is only a cop out and they know it. The truth is that the belief is not "deep seated". If it were, it would have caused them to make a decision that was in line with their value set. Rather than cheat, they would work to make sure that every possible opportunity to save money on their taxes is exploited. Bottom line, true values are deep seated, consistently applied, and they make decision making relatively easy.

When a situation comes up that is not covered by production values one way or another, a reasonable person will look for a solution that offers the least amount of cost and the largest amount of gain. If the problem is serious and requires damage control, there is a strong likelihood that the solution will meet the damage control criteria, which is to look for the least amount of pain, and the least amount of loss. The scenario most likely to result in a compromise solution would be the damage control scenario. A compromise solution insinuates that something has to be done that is not exactly right, but it comes closest to meeting the criteria (least cost – least loss) Damage control scenarios are common in manufacturing especially when costly errors are made somewhere in the system.

Should a potentially damaging outcome be reason enough to cause people to compromise their production values? I am one who thinks that production values should never be compromised. Color me radical.

Even when everything is running normally, critical decisions have to be made almost hourly. Customers change their orders with no notice and sometimes cancel. Machines break down or just don't work as they should and critical people fail to show up when they are needed the most. If you are in the production business and this never happens to you, consider yourself either extraordinary or in denial.

The unforeseen is normal in production and shouldn't be used as a reason for a breakdown in production values. In fact, I can think of very few reasons why management should even consider violating their run rules. Anything that happens can be solved within the production value set if your values are truly values and not convenient and shallow values.

A common set of values, shared by all is absolutely essential in production. Without it, you're just gambling with the success of your company. Decisions are made every day that may help one customer and hurt all the rest. In trying to solve one problem, people sometimes come up with questionable solutions that cause long-term pain. For example, where do you put that new machine? Do you look around for some open space, or do you put that machine where it will do the most good? Do you do what is right or what is convenient? To do what is right may be costly for the short term. Should you bite the bullet on cost, or should you do a cost analysis?

Let me tell you straight out that if you do a cost analysis, you probably won't choose a layout change. You will take the quick and easy route and the process will suffer because the deck is always stacked in favor of the quick solution. I've been involved in debates for layout changes and I know that it is very difficult to get senior management to approve layout changes. It is almost as difficult as getting them to pay for their neighbors mortgage. They will certainly agree that a new layout is the right thing to do … BUT. The reason for their resistance is that the payback for layout changes is always slow, even though it is certain. In addition, there are factors that never enter into the equation, such as the (potential) effect a layout change will have on quality and productivity, and the effect the layout change will have on future growth. Cost analysis only looks at the short-term effect of the decision and how long the payback will take.

Quite often in situations not involving values, reasonable people sometimes seek out a compromise that they may not think is the absolute best way to go, but, given all of the circumstances, offers them a sort of win-win scenario. On the production floor, compromising is more dangerous. A compromise solution means that there is going to be some pain, but hopefully it is minimized. On the production line, you have to find a solution that fits within an acceptable value set. When you come up with a decision that violates that value set even a little bit, the decision needs more work. Do not be so quick to accept that compromise. Don't let the cancer of compromise get its tentacles into your process.

If only we had just one set of values in our plant or business - one single set of core values to guide us in our business decision process. If only every single person from the top down shared that same value set, wouldn't it be easier to manage our process and our people? Of course it would! Everyone would know and believe in the values and would react predictably to any given situation. Upper management would arrive at the proper solution every time because their motivation would be consistent and predictable. Everybody in the plant would be looking through a single pair of eyes, values wise.

Do not think for a minute that this utopian view of production is impossible to attain, and for goodness sake, don't even consider the possibility that this sort of scenario would stifle creativity. The opposite is true. There are bad solutions, there are acceptable solutions and there are great solutions. The creativity that takes you from an acceptable solution to a great solution starts with good strong core values. The bad solutions most likely originate from a lack of solid core values. It is really as simple as that.

Think about it in a more personal way … if your own personal values are good and solid and you provide your children with sound guidance and leadership, you have a pretty good feeling that when your children find themselves at a critical decision point in their lives, they will make the same decision that you would have made. They would follow the correct path because the important decisions that they have to make would be guided by the core values that you instilled in them. If you convince them that certain acts are immoral, they will reject such acts because they **believe** that they would be doing the wrong thing, not because you told them not to do it. It is the greatest example of leadership by example.

You certainly can't watch your children all of the time, just as you can't watch your employees all of the time. Since you can't constantly watch them, you have to hope and pray that they will make the right decision when something unusual happens because you know for sure that the wrong decision can be devastating. It is always more comforting when you have set a good example for either your employees or your children. If you are true to your principles, it is noticed and emulated. Values are more likely to be adopted when the proponent of those values sets a positive example.

Every single day, in every single business, be it a manufacturing plant or a bank, something will happen that just is not covered by "rules and regulations." Every single day a decision has to be made that could have dramatic effects on the ability of the company to succeed. A wrong decision does not always have drastic effects instantly; in fact, it seldom does. It could take days, weeks, months or years to realize the negative effects of a bad decision. Process improvement efforts usually attack waste that took years to develop. Lousy layouts are a good example of decisions that come back to haunt you later.

Managers have to understand that they can't allow those bad decisions to take place on their watch. They have to establish and maintain good, sound production values and establish themselves as visible and vocal advocates of those values. They have to be such strong leaders that their people know exactly what they should do in any given scenario. They cannot be ambiguous; they have to be consistent and predictable. Furthermore, when there is a breach of the value set, management must respond quickly and strongly. I'm not suggesting that they always react in a draconian way, I'm merely stating that it should be clear to everyone that adherence to approved values is as important as compliance to any existing rule or regulation.

So where do we start when we try to decide what values to subscribe to? The answer is simple: There are certain things that are destructive to the ability of a company to succeed, and there are things that enhance a company's ability to be more competitive. We need to determine what causes the hurt and what makes things better. The values that you are about to examine haven't changed a great deal for over twenty years and they are not going to change a great deal in the next fifty years, if they change at all.

Quite often companies are of the opinion that they are different from everyone else, even those that produce what they produce. They seem to think that their situation is unique and that their questionable practices are acceptable because of the nature of their business. They couldn't be more wrong. If it was wrong for Toyota, it is equally wrong for any other business out there. In a values based manufacturing system, the only thing unique about one producer or another is simply the product they produce. The process values they maintain should be essentially the same. Right is right and wrong is wrong – period!

This section gets right into exactly what is right and what is wrong. I encourage you to think of one as the dark side and the other as truth and righteousness. That may sound a little silly to some of you but you will never really "get it" until you think in these terms. You have to get to the point that you always refuse to accept a solution that is not within your accepted value set. There can be no room for compromise. Now the good news: When you finally get to that level of belief, you will find that you will arrive at the proper solutions, and you will do so quickly. You find yourself suddenly being able to see the evil easily and instantly recognize a superior alternative. Anyone can see problems; only the true believers can see problems and solutions simultaneously. Now let's look at what is good, and what is evil.

Production Killer Number 1 – Prioritizing

The scenario goes like this: You have a great product and the demand for it is growing by leaps and bounds. At first, when the orders start coming in, there's no problem meeting the demand. Then all hell breaks loose. Your salespeople are telling you that if you could make more, they could sell more. You have been waiting for news like this and you certainly do not want to miss out on any of those orders so the first thing you do is go into an overtime mode. At first it helps, but as the orders continue to pour in, you start to fall behind on delivery lead times.

Since you're falling behind, you start releasing more work to the floor to pacify your customers who are starting to scream bloody murder. You start getting calls from customers asking where their orders are. Your brilliant answer is that their orders have been released to manufacturing. You, of course, will put pressure on your managers to speed up. You are certainly doing your part. You're applying pressure on manufacturing to improve. Why in hell won't they speed up?

Your next move is the coup de grace - the beginning of the end. In order to make your better customers happy, you decide to give their orders priority. You start out by prioritizing only late orders, and sooner than you thought was possible, you find yourself looking at the daily "hot list" and prioritizing the hottest of the hot. In the blink of an eye, you are having daily meetings of all production personnel where decisions will be made as to what will be worked first, second, and so on. In addition, you start putting pressure on the departments that can't get the work out on time.

You're on a roll now. It is certainly not your fault that your people can't hack it. You decide to go into mandatory overtime until the problem is fixed. You notice that the orders you have put emphasis on are making it through – sometimes. Then you notice that, not only are you behind on lead-time, your production numbers are getting worse. What in the world has happened and what can you do about it?

As promised, there is a culprit here and it is you! You started out pretty well with a good product and probably a good process, but you quickly lost marks after that. You made the wrong moves, and the worst move of all was to initiate prioritizing into your process. We'll talk about what you should have done step-by-step, but first let's talk about why prioritizing is such a bad practice. It is the epitome of evil!

First of all, prioritizing one order **ALWAYS** has a negative productive effect on other orders. It cannot be otherwise! When ideal conditions exist, prioritizing can have a minimal effect on production, but although the effect is minimal under ideal conditions, productivity is still adversely affected. Furthermore, the weaker the process is, the more extreme the negative effect of prioritizing is on other orders. The effect is predictable: The longer the cycle time, the harsher the effect of prioritizing, and the more prioritizing that takes place, the longer the cycle time.

When cycle time goes up, the product stays in the line longer. If the release of new product to the line remains constant or is increased, there will be a buildup of WIP on the line. As cycle time goes up, the ability to meet lead times on almost every order will become difficult or impossible. Eventually, even the orders that are prioritized will fail to get out on time. The situation only gets worse if not corrected.

Degrees of Hot: Hot, Red Hot, White Hot

The biggest problem with prioritizing is that it creates a downward spiral for production. Things do not get better, they get progressively worse. First you prioritize only the hot orders while the temperature rises on the orders that are pushed aside. Things get worse and the hot become red hot and soon you have the super white-hot orders. Before long, every single order that you release will be on the hot list eventually, if not immediately. That is what the downward spiral is all about – it just keeps getting worse and worse until you do something positive to make things better.

Those who turn to prioritizing as a way of making on-time deliveries just do not realize that they are doing something destructive. If they did, it is certain that they wouldn't do what they are doing. They do it in the name of customer service, not realizing that customer service will not improve, it will suffer. Prioritizing, as a way of meeting customer demand has never worked before, is not working now, and will never work in the future. Those that swear by it just do not realize how good they could be if they would just see the light. In other words, if they think that it is working for them, it is only because delusion has replaced reality.

Concern for the customer is something that we share, but first we have to ask ourselves which customers we are talking about. When we prioritize to meet lead times, we certainly are not concerned with all of our customers; we're only concerned with the ones that are screaming the loudest, or the ones with the largest orders. If it were true that we were concerned with ALL of our customers, we would treat them all equally. It is very easy to be fooled into thinking that the largest orders, or the customers that pay the most are the ones who should be given priority, but that sort of mentality will get a company into serious trouble. **When ALL orders are given equal priority, production can be maximized**. Controlling and avoiding prioritizing should be in everyone's value set.

Is prioritizing always wrong? Not if you're line is optimally suited to accommodate it; that is, when there is not an excessive amount of

WIP on the floor. Some companies even make additional revenue by offering their customers faster lead times at a higher cost. This option is highly attractive to companies that are struggling to meet demand themselves and require responsive suppliers. The better producers can do this without any fear of affecting other orders because they make sure that their lines are poised to deal with quick turn orders.

Quite often in these advanced companies, nothing changes when a quick turn order is released. There's no need to treat a quick turn differently because the cycle time is so low that almost nothing has to be changed to meet a quick turn. Even when the QT order moves ahead of another, the adverse affects are minimized. Some companies can prioritize 50% of their product without suffering. Others will run into serious trouble when QT orders make up only 5% of their total orders. Advanced manufacturers know that there is a limit to the percentage of quick turn orders that can be accepted and they make sure that they stay within those limits.

Again, prioritizing can be very destructive but it can be managed. If the cycle time of the line is low and the production requirements are consistent with capabilities, prioritizing can be accomplished. If you are prioritizing orders and your line is not suited for it, you probably already know this. If you know that it is killing you and you consider it an acceptable compromise, you are seriously misguided.

Customer Oriented – What Does That Really Mean?

We all think of ourselves as having concern for our customers. We proclaim to all that would listen that we are truly customer-oriented individuals. But whom are we talking about when we talk of our customers? Are we talking about the person that we speak to on the telephone when things are either going well, or not so well with deliveries? Are we talking about the company itself that we deliver product to? Or maybe, just maybe, we are talking about our own internal customers. What a concept!

IF we are really, truly concerned about our end user customers and **IF** we want to do the best for them in terms of the timely delivery of high quality products, then we are talking about our **INTERNAL CUSTOMERS**! That's right - our internal customers. If we take good care of our internal customers, we are taking good care of our external customers. If we jerk around our internal customers, we are doing the same to our external customers. If we are customer oriented, we take care to be absolutely certain that our internal customer has everything that they need to make them successful. If all work centers in a plant are customer oriented, they know that looking after their internal customer is the same as looking after their external customer.

In a production line, **everyone** is both a customer and a supplier. Every person and every work center is both a customer and a supplier. When work centers finish with a product, they then pass it on to their customer. If they are late with the deliveries or have poor quality, they will adversely affect their customer's ability to be successful. If they run out of work and all of a sudden get a huge amount of WIP too late in the day to meet their own commitments, their supplier has failed them. All work centers have to be primarily concerned with the on-time delivery of the quantity and quality of product to their customer(s). If all of the work centers shared this concern for each other, they would find themselves meeting their lead times and output requirements much more readily. Being customer oriented (Internal and External) should be in everyone's value set.

But it doesn't end with the production line. The thinking applies to support groups also. Engineering, maintenance, sales, IT and every other department have to be highly motivated to take care of their internal customer. None of us can think of only the customer that is "out there somewhere".

The Product Release Point – Responsive and Predictable (Hopefully!)

In most production plants, the product is released to the floor according to some kind of schedule. The production schedule is sometimes calculated manually, although some plants have sophisticated computer-generated planners that can do much more than just create a schedule. For example, some planners can be tied into the parts availability database and have the ability to predict lead times under various resource scenarios. The better planners are programmed to run multiple what-if scenarios because they are literally simulation modeling programs. These planners can do things that people could never accomplish in a complicated, multi-part number plant. In the right hands, and under the right conditions, these programs are really powerful tools.

That last statement is really the kicker. "In the right hands ..." is a serious qualifying statement. If the people controlling the release of product to the floor have the right production values, they will make sure that the planner is properly programmed to reflect those values. If they have questionable values, they will probably adversely affect the intent of the program by constantly overriding the schedules that the program produces. The planner is just like any other program in one respect: Garbage in – garbage out. Consider the possibilities:

1. The program could work well as planned
2. The program could be fed faulty data and put out faulty information
3. The program's output could be overrode by well meaning but misguided individuals

It cannot be overlooked that two of three of the possibilities are negative. It's the old forward pass analogy again. Only three things can happen and two of them are bad. In this case, we have to hope that a sophisticated program, properly programmed and executed will always work as planned. It will of course unless there is human intervention.

In a constantly changing manufacturing environment, it is essential for manufacturing to have control of the order release process. I'm not saying that it is necessary for them to program the planner or create the schedule; I'm saying that they should control the actual release of product to the floor. They should follow the run rules that guide the release, but they should answer **ONLY** to those run rules.

Managerial interference with the scheduled release is damaging to productivity. Management should not attempt to interfere with the release process because there is more to be considered than just a call from a good customer to alter the planned release of product to the floor. The biggest danger to a successful schedule is often a manager with an agenda. Manufacturing should always control the final release – period!

There is a very good reason for this, especially in a FCP plant: Manufacturing and only manufacturing knows what the situation is downstream from the release point. They have visibility throughout the manufacturing process and controls in place to keep WIP under control. They know the best mix of product to allow them to operate at peak efficiency. They can respond quickly to problems on the production floor. There may be breakdowns or other kinds of problems that cause the WIP to build up at the location of the problem, but they can respond to those problems quickly because they have a plan in place to deal with the problems.

Just like everything else in production, the people who control the release of product to the floor have to operate from a solid values base. The planner only creates a schedule based on the information entered into it. The situation in manufacturing at the time of release may have degraded significantly since the schedule was released. Rather than continuing to release product that piles up, manufacturing can stop the input and respond to the problem. Once the problem is fixed, the normal release can resume. It is important that product flows without restrictions and when there are restrictions, they are dealt with.

Product should be released to the floor in a predictable and responsive manner. When the line is running at optimum speed in all departments, the product should be released in optimum amounts steadily to ensure that there is always the correct amount of work on the floor. When there are problems on the floor, the amount released should be metered down to match the capability of the problem area until the problem is corrected with consideration given to WIP levels leading up to the problem area. Only manufacturing knows how much is enough and how much is too much at any given moment. Giving the release responsibility to a department that does not report to manufacturing is dangerous to say the least.

It is rare to see a production line that runs at its full capability all of the time. There's always something happening that chokes production. It could be a quality problem, a parts shortage, a lack of resources, the absence of key people, or it could be that a critical machine is down for repairs. Whatever the reason, something is always happening that inhibits production and causes WIP buildups at the pinch point unless there are controls in place.

Everyone knows that all orders are not created equally. With a large product line, some companies find that an order for one customer is very easy to produce and offers no complexities. Another order may be present some complexities but is manageable. Then there is the order that always takes much longer to produce, has a high fallout rate and just creates chaos as it moves through the line. These orders from hell are the ones that have to be prepared for and there are ways to deal with them.

The important thing to remember about the demon orders is that it is imperative that the line isn't inundated with them. They have to be metered into the line in manageable quantities rather than dumping a huge amount of them into the line and watching the brown out take place. You'll learn more about how to control this type of product later when we cover Dynabans, but for now it is important to know that these types of orders have to be metered into the line slowly and steadily so as not to cause a serious production breakdown. This type of problem could be as difficult to deal with as breakdowns and other types of production issues.

Saying that it is necessary to be responsive to the capabilities of the line is easier said than done. There has to be a reliable way of determining just when to meter down the flow and when to open the gates. Later on, when we cover Kanban and Dynaban Management, you'll learn how to determine when enough is enough and when more is needed.

Product Transport – Another Killer – Especially When It is Undisciplined

There are very few manufacturing companies that do not have to transport product from one department or work center to another. In better companies the product doesn't have to travel far to go from one place to another. In other companies, the product spends almost as much time being transported or waiting to be transported as it spends being worked on. When transport time is a significant portion of total cycle time, the plant is in trouble. Furthermore, if transport time is significant, it is probably the result of a very bad plant layout. In most cases, the plant started out with a good layout and the situation degraded as business grew and new machines were added.

Some companies think that they can minimize their transport costs by moving the product less often. That line of thinking is about as wise as spitting into a strong wind. A steady movement of product from front to rear is an absolute necessity if productivity and low cycle time is considered important. You do not get paid for product that is sitting on the floor - it has to continue to move. People that believe that it will save money to move product less often also believe that they can reduce their cycle time by getting faster transporters. Calculate in your mind the accumulation of the following:

1. Product waits to be transported
2. Product is moved from one location to the next
3. Product waits to be worked on at the new location

Hours can be added to the cycle time without any work at all being completed that adds value to the product. Since product has to be moved through the process, it is a necessary evil unless the movement is erratic and unreliable. When movement is unpredictable, it is a production killer.

When product moves through the line in an undisciplined, unpredictable and/or unreliable manner, greater amounts of "Just In Case" WIP is needed. Without it a workstation may be idle if its next shipment is late. I know of no one (in his or her right mind) who prefers a scenario where deliveries are unpredictable. People want the warm fuzzies that go with predictable shipments. There are no external customers who like to be surprised to find out that their order is late, and there are no internal customers who like it either. How would you like your department to sit there all day waiting for work only to get it right before the end of your shift? You're right, you wouldn't!

There are several reasons for the existence of unpredictable delivery of product from department to department. The most common reason is that work centers just wait until they have someone available to deliver the product. It could also be true that they are not internal customer oriented and do not care about anything but their own output. If this is the case, you can bet that productivity is something that is just dreamed about in that plant. In many plants there is no discipline that regulates product deliveries. Each department just does its own thing.

One department that I observed stacked their finished product next to the wall and waited until the product reached a height that was equal to a line that was drawn on the wall before they would move the finished work to their customer. They could care less about their internal customer's needs; they were only concerned with themselves. They didn't want to tie up one of their workers to deliver the product because they would lose output. Their whole mentality was to put all of their resources into making product and only deliver that product once or twice a day at the most.

Another very common reason for late or inconsistent deliveries is associated with the **distance** between work centers. Everyone in the production world knows that the greater the distance between work-centers, the greater the probability of late or missed deliveries. Good production people also know that transporting product over long distances has an adverse affect on product quality. In some plants there are departments that are so far apart that they are almost in another zip code. The only way that product can move from department to department is by long distance trucking.

Product transfer can be a serious problem in manufacturing. It can adversely affect quality and on-time delivery and it can certainly be costly. It is analogous to a forward pass in football. There are only three possibilities, and two of them are bad. In football, a pass can be incomplete; it can be intercepted and on the positive side, the intended person can catch it. With transport the only good thing that can happen is that the product gets to the intended department undamaged. The best thing that companies can do for themselves is to eliminate the need for product transfer if possible. If that is not possible, work to minimize the transport distances with good line layouts.

There Are Orders – And There Are Orders

Consider Scenario 1: Lately manufacturing has had difficulties with product delivery to customers. The on-time delivery percentage has dropped and customers are complaining about late orders. Senior management has raised the flag and declared that something must be done to improve the situation. Middle managers are looking for a way to determine just who is missing the boat or what is happening with the process. Meetings are held to find a solution to the problem.

The company needs to manufacture 1000 widgets daily in order to meet order requirements. Every single day, one department or another is unable to meet the numbers. The excuses vary, so it is difficult to determine just what is causing the problems. To make up for lost output, some departments have to work overtime. Pretty soon, overtime is no longer a rare occurrence - Mandatory overtime soon becomes the order of the day. For some reason, the overtime just isn't getting the job done. Management is being told that more people and machines are needed to meet requirements. This presents a dilemma because the company is already operating at a loss. The pressure to improve is intense but nobody seems to know what to do about the situation.

Now consider scenario 2: For a long time everything ran just as it should. On occasion, orders were late or there were problems that slowed production for a short period of time. Product moved through the line predictably except for occasional breakdowns. Recently however, the problems seem to be happening more frequently. Everything seems to be running just as it has in the past, with the same people in place, but the results don't seem to be consistently good lately. Something is wrong, but what is it? What has changed? If something is different, when did it happen and why didn't someone notice it?

Management is trying to determine the answers to the questions, but don't know where to start. Meetings are held to determine what the problems are and to find solutions to them. While people work to figure out the answers, the situation continues to worsen. Management decides to set quotas for each department and hold meetings to compare the actual output numbers with requirements. Management makes it clear that each department must meet its quotas. The pressure to find and fix the problem is intense.

Even though the management in the second scenario is a bit more progressive than the management in scenario 1, they both eventually end up using the same tactic: apply pressure for higher output numbers. Most managers prefer the numbers solution because they think that the problem is associated with people. The process isn't even considered because they don't know how to manage the process; they only know how to manage people.

If the process is not working well, it can be fixed by putting the pressure on the workers to improve their output numbers. What they don't know is that looking at output numbers is only the beginning. Using output numbers alone to solve problems in manufacturing is analogous to solely using pain or the lack of pain to diagnose illness or wellness. There are too many other factors that are important to consider. In manufacturing, output numbers are important, but there is much more to consider.

Problems in manufacturing do not just happen overnight and a process does not fall apart because of a single incident. A process loses its health over a period of time and usually, as a result of a series of small, seemingly insignificant acts of promulgation initiated by supervisors, team leaders and even floor engineers. Because the acts are initiated as informal decrees, upper and middle managers are usually totally unaware of their existence. The following are examples of what some very well intentioned employees can do to slowly cripple a healthy process:

- A technical lead orders workers to double count all deliveries until further notice because the supplier "shorted" them on the last delivery.
- To ensure good quality, a supervisor orders the workers doing machine setups to wait until the setup is "verified" before they can start processing product.
- A team leader orders the workers to refrain from starting new orders within one hour of the end of the shift because the workers on the relieving shift refuse to work on orders that they did not initiate. The action was taken because a worker was "burned" for rejects that were the fault of the previous shift.
- Work is stopped 30 minutes before the end of a shift to "clean up" the area.
- A supervisor orders inspectors to reject complete lots when more than two rejects are noted because there is a shortage of inspectors and multiple rejects will cause the inspectors to be overloaded.
- A floor engineer orders workers to page him and hold up work on a particular order until he arrives because he wants to personally examine the incoming product.
- Engineering takes control of a critical production tool or machine during prime production time to perform important engineering tasks that could have been accomplished during off-hours.
- A lead technician instructs the workers to set aside certain orders until she can inspect them. She is worried that the

workers are not qualified to work on certain product without her supervision.

There are many more examples like these, but I think you get the point. These were all examples of production workers being shackled, and then told to run fast. Even though the perpetrators of the preceding deeds were certainly well intentioned and convinced that they were justified in their actions, each example resulted in lost production time and additional non-value add time. For each example that resulted in lost time and output, there was an alternative solution that would not have hurt production. Too often however, what starts out as a temporary fix, turns out to be a permanent problem. The problem with promulgation in manufacturing is that the decrees or orders are based on faulty logic and values. These little cancers in manufacturing just keep slowing production while remaining hidden from upper management.

Even when pressure is brought to bear on each department to meet requirements, the workers in those departments are forced to follow both the written **and** the unwritten procedures. As far as management is concerned, only the written procedures have to be followed. As far as the workers are concerned, they have to do what they are told to do as well as what they are officially required to do. If the extra work that they have to do causes them to miss their output numbers, it certainly isn't their fault, but they often end up getting blamed anyway.

How can this mess be fixed? First of all, there has to be another way of determining the individual capabilities of each department. Just looking at the output of each department as a report card for productivity is essentially the same as looking at total sales as the sole report card for profitability. One indicator is not enough to evaluate the productivity of a department. So, if one indicator is not enough, then what others should be used and why? Looking at output is intuitive to manufacturing people. When output is good, everything is rosy (or so they think). When the situation turns sour, what else is there to do but put pressure upon the department whose numbers have sunk.

This brand of logic (numbers) is pervasive among manufacturing people. The logic is simple but straightforward and is the logic that is preferred by power oriented managers because it allows them to use managerial pressure to achieve results. There is no skill required to manage a numbers situation – simply put the appropriate amount of pressure on the allegedly guilty and eventually they will find a way to improve their situation. If they are unable to improve, the next solution is equally simple: find someone that **can** get the job done.

Setup Reduction vs. Large Batch Sizes

When machines started getting bigger, faster and more technologically complex, it became clear to production people that the amount of time that it took to "prepare" the machine for a different group of widgets increased dramatically, another way of saying that Setups were taking too long to complete and were seriously affecting output. There had to be a way of dealing with the long setup times. Engineering came up with a solution: Increase Batch sizes!

By increasing the batch sizes significantly (10X) it was possible to produce more **widgets** because you did not have to spend a significant amount of time doing setups. The theory sounds fine until you consider that the purveyors of large batch sizes were working with the wrong variable. They "assumed" that machine setup time was a fixed number. The only variable that they chose to work with was the size of the production batch. The theory is simple: if a setup takes 4 to 5 hours to complete, increase the batch size to a number large enough to take the sting out of the lost production time due to the setup.

Instead of running 100 widgets and then doing a 4 hour setup, run 1000 before you do that 4 hour setup. Better still, run 5,000 of them hummers and you will save even more setup time. You can really be productive then. Batch sizes across America went up dramatically. In Japan, Shingo was shaking his head at the madness because he knew that customers liked variety. He knew that all automobiles were not black like they were in the days of the Model T. There was a demand for a variety of colors and the public was demanding even more options.

Batching theory is great on the surface, but it is impossible to run one product type for a full week before you switch over to another type. It is important to ship some of each product type each and every day. If you have to do 10 white cars and then do a setup for 10 red cars, and so on until you cover all of the colors, you will not get many cars completed because you will be spending all of your time doing setups. Batching didn't work because of the demand for variety.

Shingo asserted that the Americans were concentrating on the wrong variable. They were treating the setup time itself like a sacred cow. He rightfully believed that the setup time itself could be reduced so significantly that batch sizes could be kept at a minimum. He was so right! Shingo showed how setups could go from hours to minutes by concentrating on the details of the setup and in effect, simplifying the setup and removing the waste

Shingo's discovery also explained how this "magic" could be realized in any production plant. His solution was not complicated like brain surgery, but it certainly was brilliant in its simplicity. Here is a perfect example of a **Right Hand Simple** solution. As a solution, it is both simple and inexpensive to implement and it pays off with big dividends. The beauty of this solution is that it not only increases output, it also works to cut cycle time by reducing the need for large amounts of WIP. Large batches only gives you more output. As a solution it is left hand simple. It may give you results if you have only one product type, but how many companies have that? Even so, the cycle time for the product in a large batch scenario is huge compared to what it would be if setup reduction were used.

Setup reduction should be a way of productive life for every company and should be part of a continuous improvement effort. It is not just a one-time effort because there has to be a great deal of follow-up involved in setup reduction. Once the setup times are reduced, each individual setup must be audited as insurance against recidivistic behavior. In addition, the initial effort may have uncovered other opportunities that can only be acted on over time, such as machine modification that would facilitate and further reduce total setup time.

The FCP way of implementing a setup reduction effort will be covered later on. Right now it is important to point out that when setups create problems by stifling production output, your value set should lead you towards setup reduction and not towards an increase of batch sizes. All that the latter will do for you is to increase your cycle time. You can't just think about numbers, because if you do, you are taking the wrong road. The road you want to take is the road to recovery, not the road to ruin.

Chapter 2 – Modern Production Tools

There is no doubt that there has to be ways to determine how well a production line is doing. I have been encouraging (nagging is better) you to find ways to measure the productivity of your process rather than concentrate on the daily output of product by your employees. My reasoning is simple: Concentrating on employee performance does nothing more than blind you to the important measurements that not only lead you to conclusions, but to solutions.

Product output is certainly important, but it is not the most important metric. It is one of three. Working to open constrictions and improve output numbers is certainly a very important goal of any manufacturing company. It tells you something, but it doesn't tell you enough. It is analogous to a reliance on the level of the gas gauge to determine the efficiency of your automobile engine. You just need to know more.

If You __Must__ Measure Something - Measure This!

There are three equations that help management to determine just how healthy their processes are. These equations are best used as departmental or work center measuring tools rather than the complete factory. They are:

1. Little's Law – Used to measure the Cycle Time as a product of WIP and Output.
2. MCE – Used to measure the ratio of Value Add Time to Total Cycle Time.
3. X Factor – Used to measure the ratio of Raw Process Time to Total Cycle Time.

Little's Law is clearly the most important measurement in production. MCE and X Factor cannot be determined until Cycle Time is calculated. Cycle Time is the Big Kahuna – the mother of all measurements. But as you will learn, although it is important to have low cycle time, how you get it is even more important. Later you will discover that keeping low cycle time day after day is the greatest challenge faced by production. Achieving and keeping a low cycle time and consistently good output is certainly attainable, but it requires a sustained effort.

Little's Law – Cycle Time = WIP / Output

All things being equal, the company that can reliably produce a quality product in less time than its competitors will enjoy a decisive competitive edge. Many companies could sell more of their product and gain market share if only they could consistently ship their orders on time and in the proper quantities. Every company hesitates to turn down orders even though they are not certain that they can meet the promised lead times or required quantities. When a company takes an order from a customer and promises to deliver the product on a certain date, the customer is sure to be very disappointed when there are delays. When a producer makes promises and can't keep those promises, its customers eventually start looking around for a more reliable company to do business with. In the tough world of production, the future belongs to the company that can exploit opportunity when it exists and maintain its edge over the competition. First, let's define some of the terms that we'll be using.

Product can be defined as an automobile, a refrigerator, a printed circuit board, a newspaper, a sofa, a television set, a computer, or a widget. It is anything that a company produces that its customers are willing to pay for and take possession of. Product could be a refrigerator, a sofa, or a loaf of bread. It could also be a completed loan, a finished home, a bank transaction, a maintenance service or a pharmaceutical prescription. Product is what the end customer wants delivered on time.

Work In Process (WIP) is defined as product that is trapped within the process, and therefore is often referred to as Trapped **WIP**. Trapped **WIP** is every bit of product that exists in the line from the first process step in the line to the last. Trapped WIP could be measured within the complete line or within one department or work center. WIP is product trapped in the process. The customer won't pay for Trapped WIP; they will only pay for finished product delivered to them.

Product **Output** is defined as the average amount of product that the line, the department or the work center is able to ship to the customer. **Output** could be measured hourly, daily, weekly or

149

monthly, but as you will see when we cover TAKT Management, the preferred period of time to measure output is hourly.

The **Cycle Time** of the product is the total time it takes to process the product from the moment that the product is introduced to the line to the moment that the last process step has been completed. The Cycle Time of a department is the total time it takes that department to process the product from the moment that the product is delivered to the department to the moment that the product is ready for delivery to the next department or work center.

Lead Time is the total time that it takes to get the product to the customer from the time that the order is received until the time that the customer receives the product.

The "Unwritten Messages in Little's Law

Although Little's Law is an equation, there are some complex messages that are not immediately apparent to the untrained eye. The "unwritten" message contained within Little's equation is that the three most important elements of production are: product **Cycle Time**, **WIP** and **Output**, and that **the way** that **Cycle Time** is improved is of paramount importance. Once the significance of reduced cycle time is acknowledged, it becomes clear by examining the equation that there are only three ways that lower **Cycle Time** can be accomplished:

1. Reduce excess **WIP**
2. Improve **Output**.
3. Improve **Output** <u>AND</u> Reduce excess **WIP**

You can take this to the bank: The proper way to achieve the most optimum **Cycle Time** for a process is to maintain the least amount of **WIP** necessary to meet changing process needs while working to simplify the process and improving **Output** capability. Low **Cycle Time**, high **Output** and low **WIP** are the ingredients needed for a world-class process.

A great deal of care must be exercised when the tactic employed for reducing cycle time is simply to reduce the amount of **WIP** in the process. In fact, a common mistake made by inexperienced producers is to indiscriminately reduce the amount of WIP without determining just how much is just enough, too much and too little. If an excessive amount of **WIP** is removed, the **Output** capability will be limited. If too little is removed, optimum cycle time cannot be achieved. There must always be an appropriate amount of WIP present in the process to meet changing needs.

Improving output is the best way to reduce cycle time and the best way to improve output is to simplify the process by removing waste from it. The best description of a world-class production process is to say that it is devoid of wasteful process steps. A company cannot provide their final customer with the required amount of product at the appropriate time if it does not maintain that same standard within its own production process. The internal customers must receive the same consideration as the external customer with respect to quality, quantity and timeliness of delivery.

It is very important that the **amount** of product entering the process and the **frequency** of product entering the process is adequate to meet the output requirements. The amount of product at the release point has to equal the output requirements PLUS the amount of product that will be lost during production due to quality issues and damage. Managing the input to the line is a challenging, and underrated job. Consideration has to be given to product mix, batch sizes, customer orders and appropriate quantities.

Controlling WIP within the line once it has been released is best accomplished through Kanban Management, which is covered in more detail later. When Kanban management is in place, a flag is raised when there are WIP buildups within the line. When there is a line stoppage, continuing to release product to the line is the wrong thing to do, but there has to be a way of sending a stop signal. Of course, the complete line does not stop at once, the buildup stops shipment to the department where the buildup began and it moves from that point towards the front of the line all the way to the initial input point. Kanban Management is very effective at guarding against excessive WIP buildups.

If the output capability of a department is exceeded by the input rate, a buildup of WIP will occur over time. This buildup of WIP will cause the cycle time of the product that is trapped in the process to increase. The only solutions to this dilemma are to increase the amount of output to equal the input or to restrict the input. Whenever the cycle time of one department goes up, the cycle time of the factory goes up. Understanding where the problem is and working to correct that problem is the reason that Fast Cycle Production prefers looking at the cycle time of departments as well as the cycle time of the complete factory.

Little's equation states that, in order to determine cycle time, you divide the total amount of units trapped within the line by the total output of the line. The equation is:

CT = WIP/Output

Let's calculate the cycle time for a typical department using hourly output and present trapped WIP.

Average WIP = 1000 Widgets
Average Output = 100 Widgets per hour.
CT = 1000/100 = 10 hours

Remember that the input to the line is hourly, meaning that the amount of WIP will remain at an average of 1000 and the output may go up or down, but will average 100 per hour. If this assumption remains true, cycle time will remain constant unless action is taken to change either the total amount of WIP or the hourly output. If the output increases, the cycle time will go down. If WIP is reduced, the cycle time will go down. If output goes down and/or WIP increases, cycle time will increase. Nothing else is possible.

Going back to the example, to calculate cycle time for this line we divide 1000 by 100 giving us a cycle time of 10 hours. That means that on average, each Widget will spend 10 hours in the department. It may take only one hour of work to complete all of the production tasks, but each Widget will remain in that department for 10 hours. A reasonable person might ask why it takes 10 hours to receive and ship product when the total amount of work that is put into the product amounts to only one hour. The reasonable answer to that question is that it should not take that long, but in most companies and in most departments; a ten to one ratio is quite common. As you will soon discover, the more productive companies have a ratio of five to one **or less.**

What is good cycle time and what is excessive cycle time? What is considered "World Class" cycle time, what is considered fair and what is considered poor? Little's Law alone will not answer those questions. To understand just how well a process compares with other processes, other equations are required that work in tandem with Little's Law. Here are a couple of equations that begin with cycle time to determine how the line is performing.

The X Factor (X_f = CT/RPT)

X Factor tells you how many times your Raw Process Time can be divided into Cycle Time. If Cycle Time is 10 hours and Raw Process Time is 1 hour, the X Factor is 10.

X_f = **CT/RPT** (CT = 10, RPT = 1) (10/1 = 10).

This very important equation actually originated with CFM and was usually used with departments that were primarily machine oriented. RPT was determined through a IE study of the machine(s) in the department and only the machine time or the actual time that the machine took to process various types of product were used to determine RPT. The type of department that used X Factor in the CFM world would be a department that processed product primarily by machine, not by hand. Since it was so difficult to accurately determine RPT, X Factor was not widely used.

FCP has a much different view of the usefulness of X Factor and the determination of RPT. On a line practicing FCP, RPT is defined as **any** process step completed sequentially that is officially required by management. Unlike **MCE**, which will be covered next, X Factor does not differentiate between tasks that are value add and tasks that are non-value add. If the task is officially required and it is performed sequentially, it is included in calculating Raw Process Time.

There are very good reasons for including all required sequential process steps in measuring X Factor. First, a tool that requires complicated engineering studies before it can be used is going to be considered a "one-shot equation". Just do it once and then forget about it. If that were the case, it would be a worthless equation. As it is, it can be a very useful equation that tells a department that there is either too much trapped WIIP or the product is just lying around too much. Either way, something is wrong that needs to be fixed.

So, what do you want your X Factor to be? Well, world-class manufacturers have an X Factor from 3 to 5! This means that their RPT is only one third to one fifth of their cycle time. This may not sound so impressive to some of you unless you understand that some of the larger American companies around have X Factors of 10 or more. What may be more significant is the fact that the lower the X Factor, the higher the probability of delivering product to the end customer on time. Adding to that is the fact that, as you will soon see, a low X Factor is needed in order to prioritize some of your orders.

The usefulness of solving for X Factor is not just to come up with a number, it is to determine whether the Cycle Time is acceptable or not. An X Factor of 10, when compared with a world-class ratio of three means that there is a great deal of room for improvement. The usefulness of X Factor doesn't just end with the discovery that it is good or lousy. A high X Factor is also a sign that the process cannot respond positively to urgent demands for increased output or the prioritizing of orders.

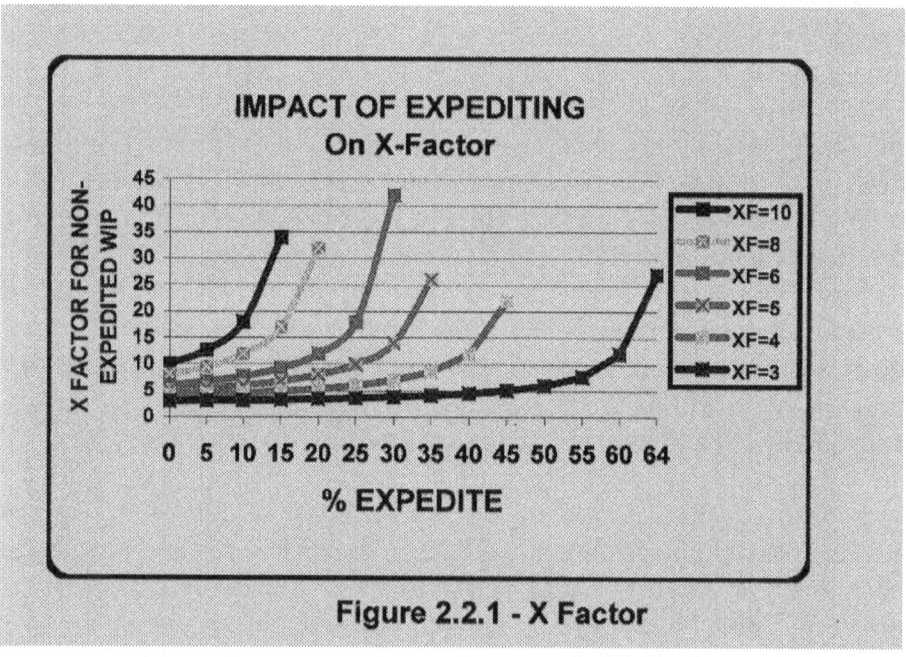

Figure 2.2.1 - X Factor

World class producers with X Factors of 5 or less throughout the plant, can sell quick turn (high priority) orders without concerns about normal production problems, such as machine breakdowns, part shortages and resource issues. If these kinds of problems happen, the output capability of the process will not be highly affected because that process is under control and product flows smoothly throughout the line. Vulnerability to problems is directly related to high X Factor values.

Figure 2.2.1 dramatically illustrates the effect that prioritizing has on X Factor. If a plant with an X Factor of 10 prioritizes 20% of its orders, its X Factor increases to about 34, an increase of over 300%! A plant with an X Factor of 3 can prioritize 50% of its orders and its X Factor will only increase to 5, a very acceptable number. When a plant or department has a high X Factor, every production problem has the potential of becoming a disaster. The chart clearly shows that prioritizing is not a problem when the X Factor is 5 or less. Any X Factor number in the double-digit area has significant problems with even a small amount of prioritizing.

Anyone who has been in the production business for over a week knows that the occurrence of unforeseen problems is a way of life. The question is not whether problems will happen; the question is **when** those problems will happen. Machines break down, mistakes are made and customers change their demands. Preparing for the eventuality of problems is smart business. The best way that a plant can prepare itself is to establish and maintain a low X Factor. Problems don't become disasters when X Factors are low.

Manufacturing Cycle Efficiency (MCE = VA/CT)

When it comes to evaluating a potential solution, MCE is a tool that is unbeatable in its usefulness. When cycle time is reduced significantly **AND** MCE is increased, the solution gets an "A". If cycle time is reduced and MCE is not increased, the solution definitely needs more work.

MCE (MCE = VA/CT) is used to determine the percentage of Value Add Time (VA) to Cycle Time (CT). If the total amount of Value Add time for a process is 1 hour and the total cycle time for that process is 10 hours, Value Add Time is 10% of Cycle Time. It is preferable to present MCE as a whole number because whole numbers are easier to deal with. To accomplish this, solve the equation and multiply the answer by 100 (MCE = 1/10 = .10 * 100 = 10).

FCP uses MCE in determining how much waste is in a process and is traditionally used to show just how much time the product spends traveling, sitting around and working on something other than the product, such as counting, inspecting, or paperwork, unless of course you are in the counting, inspecting, or paperwork business in your department. If your responsibility is to produce widgets, then only the time that you spend making the product more acceptable to your customer is considered Value Add. If your department is responsible for inspections, then only the time that you spend inspecting is Value Add.

Was The Solution a Good One? How do you know?

All solutions are not created equal. Some solutions will yield great results both long term and short term and at the same time be inexpensive to implement. Other solutions may yield excellent results but are expensive to implement. An example of the latter is the purchase of an expensive machine that puts out more widgets per hour. It sounds like the solution is sound until it is discovered that the machine can put out more widgets than the rest of the process can handle. You don't need a machine that puts out 1000 widgets an hour when all you will ever need is 100 an hour.

Going further with this scenario, the more feasible solution would be to find a way to increase output through process simplification first, rather than go straight to an expensive capital purchase. By removing wasteful process steps, both the output and the MCE would be improved significantly. Cycle time, output rate and MCE improves. Now THAT is a great solution! Along with the improved performance, you have a solution that is thousands or tens of thousands of dollars less expensive than the purchase alternative

If, after simplifying the process you still are not making enough widgets, then a capital purchase may be the thing to do. But now, instead of buying that big, honking machine that you would have bought before, you're now looking at buying just what you need. The smaller machine will probably have a smaller footprint and be easier to install in a useful space. The point being made here is that **IF** Cycle Time is improved **AND** MCE is improved, the solution is very likely a good one. On the other hand, if Cycle Time is improved and MCE **is not** improved, the solution needs more work. Keep working on the solution until both CT and MCE is improved.

MCE yields the ratio of product value add to total cycle time as a percentage. An MCE of .40 means that 40% of the work completed has contributed directly to the value of the product from a customer perspective. If, for example, you are building a refrigerator from scratch, adding value to the product, which in this case is the refrigerator, means that you are literally adding something to the product (motor, coolant, fan, frame) that brings the refrigerator closer to completion in the eyes of the customer. When the refrigerator is completely built and ready for use, all of the value add steps have been completed.

Care must be taken when calculating total value add time. Just because a task is value add, does not mean that it does not contain waste. A simple task such as adding resistors to a circuit board could be given a value add time of five minutes, when in fact, the task realistically takes seconds to complete. Total value add time can easily be both overstated and understated. In each case, the result would be an incorrect MCE value. The old rule of "Garbage in – garbage out" really applies in this case. When value add time is calculated, it must be calculated correctly or it loses its usefulness. When calculated correctly, it becomes a very powerful productivity analysis tool that could make a big difference in a company's decision process.

When working to improve the ability of a process to be productive, the MCE equation provides a way of evaluating potential solutions in a very realistic way. The best way to improve cycle time is to simplify the process by removing waste. When wasteful process steps are removed from the process, the percentage of value add steps increases. The better the solution, the higher the increase in the percentage of value add.

Calculating cycle time and MCE is best accomplished through simulation modeling, however it can also be accomplished by completing a very detailed process mapping effort that utilizes flowcharting and data collection. In a Fast Cycle Production project, process mapping always precludes a simulation modeling effort so both methods are utilized. If simulation modeling is not used, Little's Law can be used to calculate cycle time and MCE can be calculated manually. It is not as accurate as what can be accomplished through simulation modeling, but it will provide you with a ballpark figure.

All in all, X Factor and MCE are excellent ways to determine the health of a process. X Factor can be used daily, while MCE is an excellent tool to use in a process improvement project and/or a process mapping effort. MCE is used in tandem with CT and gives a report card grade to the potential solution to a problem. X Factor tells you how vulnerable your process is to quality problems hitting the line or an increase of prioritization. In either case you answer two, very important questions: How much waste do you have and what is the ratio of CT to RPT? Just knowing the answers to these questions helps you to determine whether you have a problem or not, and what you need to do about it. These two, combined with Little's Law, are great tools for both manufacturing and non-manufacturing businesses.

The Master Brewer Concept

Did you ever notice how a can of beer (name your brand) always tastes the same day after day, year after year? Whatever type of beer you prefer always tastes the same because the master brewer makes sure that the purity of the formula is maintained, even though there are variables that must be taken into consideration and compensated for. There is a famous whiskey brewer that brags that they have had less than a dozen master distillers in their hundred years or so of existence.

What is the importance of the master brewer? The answer is quite simple: there is no other person that knows how to brew beer exactly to specifications, achieving perfect results batch after batch. They follow a time proven recipe right down to the nit and they do so time after time. Thanks to people like me, millions (I think) of gallons of beer are consumed daily throughout the world. This must mean that there must be one heck of a lot of master brewers out there, right? If you think that each company had a bunch of master brewers, you are wrong. You would be right if you think that future master brewers are constantly being groomed, which is a sneaky way of saying that they are going through an apprenticeship program under the tutelage of the master brewers themselves.

Think of the master brewer as a person that knows every single trick of the trade that there is. Their job is both an art and a science to them. They do not have to brew every single batch personally, but they make sure that it is done correctly every single time. They accomplish this by applying their skills through the efforts of their apprentices. The apprentice is the person that does the work under the close scrutiny of the master. The less trained the apprentices are, the more closely they are guided by the master. As the apprentice becomes more skilled, the master backs off appropriately, but still maintains close scrutiny where the attention is required until the apprentice becomes a master.

Consider the hiring of a brand new apprentice who knows the mechanics of brewing, but not the intimate details required to produce flawless batches. He or she is not just allowed to go ahead and brew beer with the hope that everything will turn out well. That is not going to happen because the quality of the brew has to be flawless. Still, the new hire has to be trained. So how do you train someone and at the same time make sure that the job is done right every single time? Well, that is what the master brewer training concept is all about. Is the master brewer concept applicable to industries other than beer brewing? You bet it is!

There are some jobs in production companies that take a very long time to learn and where mistakes made by the operator can seriously affect quality and quantity. New people are often given formal training and then turned over to another, more experienced operator to train. During this training period, the output from the trainee and the trainer is adversely affected because of the training process. If you use a master brewer properly, you never have to worry about lost production or questionable quality with new people. The master brewer concept is not a one-on-one training concept; it is a one (master) on many (trainees) training situation.

The Master Brewer Concept offers a very workable solution when one or more of the following situations exist:

- The job in question requires a great deal of training and/or experience.
- The job in question requires a great deal of skill and discipline.
- Turnover is high. By the time they are trained, they are gone.
- High percentage of temporary workers in skilled positions. These "temps" can only work for a fixed amount of months, and then they are replaced.
- "Tricks of the trade" knowledge is an important part of the job.
- Output must be maintained even with new people.

- Quality must always be near perfect.

If your business can relate to more than one of the above conditions, using the Master Brewer Concept in your business may be a good fit. There are some prerequisites that have to be met, however. First and foremost is to determine if you have any qualified candidates for the job of Master Brewer. Finding good master brewer candidates is easier said than done. Finding several in one department is very difficult.

The person(s) that you choose should be more than just a step ahead of their peers in knowledge and experience; they should be unquestionably at the top of their class. They should also be well respected by management and peers, and have the most important attribute of all: the willingness to pass their skills on to others. They have to be competent and willing teachers. If they derive their self worth from being individual contributors, they are not likely to be willing to share their knowledge. They would be more likely to want to keep their secrets to themselves. They derive their self worth from being the best, not from being a teacher.

Again, the Master Brewers **direct** and instruct, they do not do the work. Since the Master Brewer does not actually perform the work, he or she closely monitors the work of a group of people, one at a time. They are not supervisors, they are skilled workers applying their skills by guiding, helping and teaching not only the technical skills, but the types of skills that can't be written on paper: they teach experience. Master Brewers acquire their own skills over a long period of time, but can teach those skills to others quickly. They keep their apprentices from making mistakes that they would have made without the guidance of the Master Brewer. Apprentices will avoid costly mistakes and gain valuable experience through the master's tutelage.

The span of control for the Master Brewer should be large enough to effectively utilize their abilities, but small enough so that they can be where they are needed when they are needed. A worker should not have to wait for the master at a critical moment when they need guidance. The master should make sure that each worker is getting the appropriate amount of attention for his or her level of skill. Since each individual worker's learning curve varies, the master should accelerate the training when a person absorbs the material quickly and gives additional scrutiny to those that are struggling a bit. In case the master is absent from work and there is no qualified replacement, it is quite often best to pre-determine who the master's replacement will be to avoid confusion and loss of productivity.

There are some hard, fast rules that should be adhered to in using Master Brewers. Notice that there is no hard fast guidance for span of control, and for a very good reason: the proper span of control will vary widely from business to business. There is no specific algorithm that determines proper span of control, but it should be large enough to keep the Master Brewer busy, and small enough to ensure that each person gets the guidance they need, when they need it.

Master Brewer Rules

- Responsibility for quality still belongs to the worker. Do not hold the master personally responsible for quality unless the guidance given to the worker was faulty. Each person should maintain responsibility for his or her own work because it sets the proper precedent.
- The Master Brewer should always work to develop his or her replacement. Talented protégé's can act as a temporary replacement for the master when he or she is missing
- The master is never assigned his or her own machine or work station. This would cause their attention to be diverted to their own work and not to their apprentices.
- The master cannot be a manager or supervisor. They should not be given the responsibility of reacting or dealing with issues that are normally covered by management. They should not have to deal with disciplinary issues. Their responsibilities should be clear-cut: Teach, train and mentor.

The key to effectively using the Master Brewer concept is to make sure that it does not turn into something that you do not want it to be. This type of training system has to be set up properly and monitored constantly to ensure that it does not mutate.

If used effectively, the Master Brewer concept should accelerate the training of new workers. Over time, it should be instrumental in building an impressive number of truly skilled technicians. It will help the trainees to learn all of the little tricks of the trade in a fraction of the time that it would normally take. There are very few training programs that can accomplish that. This concept is not suited for every company, but it is certainly well suited for many.

Single Unit Flow

When you have a product that requires a series of interdependent sequential tasks, it is always advantageous to complete these individual tasks in a manner much like a conveyor. As the conveyor moves along, the widget is worked on step by step. Each individual unit is completed and ready to ship when it reaches the end of the conveyor, so to speak.

If product cycle time is considered important, and it should be, then it is important that your process promotes lower cycle time. Whenever batching takes place between process steps, the first widget finished waits until the last widget to be finished in order to move on to the next station. Sometimes it is advantageous for the product to be batched, but too often batching takes place regardless of whether it is advantageous or not.

Two things happen when batching takes place ... the amount of WIP on the production floor increases and the product cycle time goes up. It doesn't just go up a little; cycle time could go up considerably depending upon how much product is stored before it is transported. The batch size between process steps may be 15, but it also may be difficult to transport every single batch. It is not uncommon for transport lot sizes to be 5 or more batches. With transport lot sizes that large, cycle time is usually a problem.

In many cases, cycle time could be reduced more than significantly with single unit flow. Complex inspections, in-circuit test and repair and similar processes are prime candidates for Single Unit Flow. Consider the side-by-side comparison of a Single Unit Flow and Batch processes shown in Figure 2.2.2. Although the process steps are identical and each get the job done, the similarity ends there. In the SUF process, the first widget is completed after 4 time units. The complete lot is completed after 18 time units. It takes 46 time units to complete the first unit and 60 time units to complete the batch. Cycle time for the SUF process is less that one third that of the batch process.

The key to SUF is not that output is improved; it is the fact that the SUF process requires much less WIP to produce at least the same amount of product as the batch process. That alone makes it the better alternative. In order to have a continuous flow of product, it would require one batch in front of each process step since the batches only move as a unit. The SUF process needs one batch every 15 time units so the deliveries to the SUF process can be continuous with one-fourth the amount of WIP.

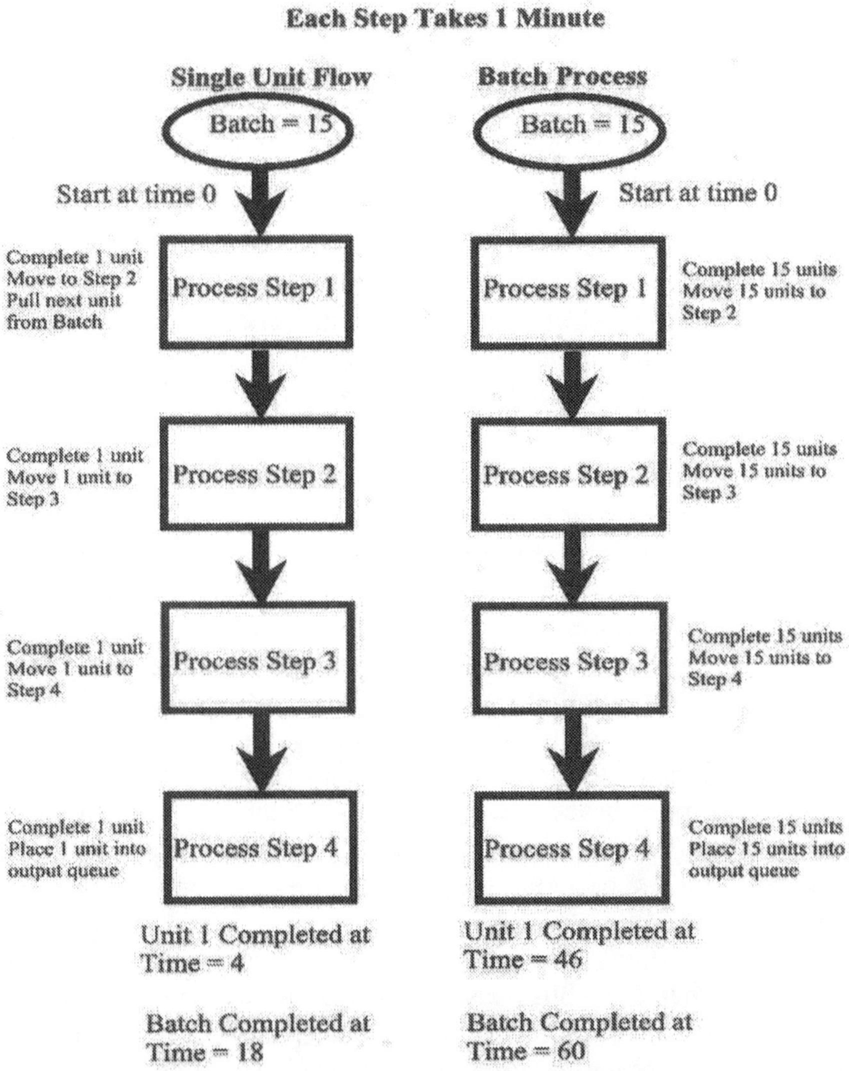

Each Step Takes 1 Minute

Single Unit Flow	Batch Process

Batch = 15　　　　Batch = 15

Start at time 0　　　　　　Start at time 0

Complete 1 unit Move to Step 2 Pull next unit from Batch	**Process Step 1**	**Process Step 1**	Complete 15 units Move 15 units to Step 2
Complete 1 unit Move 1 unit to Step 3	**Process Step 2**	**Process Step 2**	Complete 15 units Move 15 units to Step 3
Complete 1 unit Move 1 unit to Step 4	**Process Step 3**	**Process Step 3**	Complete 15 units Move 15 units to Step 4
Complete 1 unit Place 1 unit into output queue	**Process Step 4**	**Process Step 4**	Complete 15 units Place 15 units into output queue

Unit 1 Completed at Time = 4　　　　Unit 1 Completed at Time = 46

Batch Completed at Time = 18　　　　Batch Completed at Time = 60

Figure 2.2.2 – Single Unit Flow vs. Batching

Little's Law doesn't lie; the SUF process is far superior to the batch process. As noted, it requires much less WIP so the cycle time is significantly improved, but there are some intangible benefits in the SUF process that are not readily apparent.

First off, if there should be a quality problem or an engineering change that either did, or did not require rework, the impact would be much lighter for the SUF team. The batch team would have at least 60 units to deal with – four times as much as the SUF team. The significant of these intangibles should not be taken lightly, nor should the possibility of such an occurrence be considered unlikely.

Such a scenario is not only probable - it is an absolute certainty. Quality problems and engineering change orders hit without warning, and when they do, the result is **always** lost output. It cannot be otherwise! Employees are pulled from their jobs to do the rework, and they certainly cannot work on new units while they are busy doing rework. If you add shift changes, absent workers and similar daily occurrences, the negatives are magnified. Batching may be necessary in some cases, but if it is not absolutely necessary, stay away from it!

The Farm Moves to Manufacturing

When it comes to layouts, there are two philosophies that prevail in manufacturing: Farm and Group Technology. Farm technology is analogous to planting nothing but corn in one field, beans in another field and beets in yet another field. Each farm contains only one type of product. In production, that would equate to having one department for electronic test, another for analysis, and yet another for repair. If Group technology was employed; in each field you would see cells that each included a test, analysis and repair function. Each cell would contain a complete process, instead of a singular function. Single Unit Flow thrives in a Group Technology environment and Batch processes are synonymous with processes employing Farm Technology.

In a production plant, it is normal for a company to use group technology when their business is young. After all, they are small and have no need for multiple machines. Once the orders come in hard and fast, and more capital equipment is needed to meet the demand - in comes Farm Technology. In some cases, it is a good move because of changing product types and other reasons. When the company really grows, it is sometimes too difficult to just create another line identical to the one that they started out with. In some cases it is just more practical to convert.

I am not saying that every company should use group technology, because when it comes to production, neither technology is suitable for all processes. However, applying Farm Technology in every single department may be a serious mistake. Some companies revert to farm technology in situations that are not suited for it resulting in a loss of productivity, increased process waste and increased cycle time. One example of this is electronic test, analysis and repair processes of printed circuit boards (PCB's). This type of process is very likely to be converted to Farm Technology in growing or large plants. Separating test and repair process steps is unwise to say the least. They function best when they are co-located.

If you took a batch of 15 PCB's and tested each one, analyzed those that failed the test and repaired those that required repair, when you were finished with the 15 you would be completely finished. The product would be ready to ship. Looking at a farm technology setup, you would test all 15 PCB's in the test department; transported all 15 to the analysis department where those that failed would be evaluated. Then you would send all 15 to the repair department where those that required repair would be dealt with. The repair department would transport the same 15 PCB's back to the test department where those that failed would be re-tested. Finally, after all that, the 15 PCB's would be ready to ship to the next department.

Does the farm technology sound ludicrous in that situation? Well if it does, get ready for a shock … it's very common for test, analysis and repair to be separated! Companies grow, and some of them just don't know how to grow properly. There is seldom any meaningful thought given to growth, especially fast growth. It is most likely that when new test machines are needed, the most likely scenario is that the new machines will be located in the same room. Since there is no room for anything other than test machines, the analysis and repair functions are moved to another room.

It seldom matters that the process that is moved may be located some distance from its original location because the perceived need is not to keep the processes together, but to just find a place to move to. It is seldom that any thought is given to creating a new line identical to the original. The mentality quite often is just to put all of the test machines together, which often results in the separation of the test and repair processes.

Floor Layout – Only Important When the Plant is Brand New

The layout of a plant and/or work center can have a great deal of effect on productivity. Look at the possibilities: Some plants have poor production capability and a good layout. Other plants do a pretty good job of producing and have a lousy layout. The most likely scenario is a plant that has very poor productivity and a lousy layout. The last, and most rare, is the plant that is a world-class producer with a superior layout. But whatever the situation, it is likely that the plant started out with a good layout and as the business grew and the process changed, concerns for maintaining a proper layout was given a low priority.

In the first scenario, (good layout – poor productivity) the problem is directly associated with a faulty process, but the prognosis is certainly a good one. If this company were to analyze their process and remove the waste from it, their productivity increase could be very significant. In the second scenario (good productivity – lousy layout) the management of this company is strong and disciplined and probably has good production values. It has to be so because they are able to overcome the negatives that are normally associated with a bad layout. In other words, they are able to be successful IN SPITE OF their layout.

The next example (poor layout – poor productivity) is the exact opposite of the final example (great producer – great layout) in every way that is important. On one side you have a company that is struggling with their process, and their efforts are beaten down even further by the fact that their layout restricts their productivity even more. Companies that find themselves in this situation don't last long in this competitive world. Their only hope would be to change both situations quickly if they want to stay in business.

On the other hand, if they are competing with the company that has everything: a great process and a great plant layout, they might well pray for divine intervention because a world class producer like that will literally eat their lunch. Of course, even a world-class company can go downhill if they experience fast growth and do not manage it well. Production is a tough game to be in. If you want to be in that game, you have to take advantage of every weapon at your disposal. As noted, it is one thing to start with a great layout and it is quite another to go through growth and still maintain a great layout – which is why some companies start out well and their production capability deteriorates over time.

Integrated Work Teams – A Great Tandem for Single Unit Flow

An integrated work team is defined as a team of individuals that can all perform every single duty required by the team. Using the example of an In-Circuit Test (ICT) team, where the required duties are as follows, each of which is performed by a different person:

1. Setup and Operate the test machine
2. Troubleshoot and/or verify failures
3. Repair defects

An integrated work team can work each of the positions required, thus making them more flexible and less vulnerable to output loss when one of the team members is absent either temporarily or for the day. Integrated work teams always average higher output than teams that are not integrated because of their additional flexibility.

Teams ALWAYS perform better than individuals, and talented teams perform much better than talented individuals. Even when the integrated work team has less talent and experience than a team of individuals that specialize in one job, they realize more output with equal quality. Integrated work teams are more productive because they can do something that non-integrated teams cannot do – they can **FOLD!**

Folding – Also Known As: Helping Out Wherever The Need Exists

Folding is defined as moving from the position that you are working and either helping one of the team members with a temporary work buildup, or filling in for them while they are gone, or are falling behind. Folding is a powerful work technique that can have a very positive and dramatic effect on productivity. When an integrated work team practices folding, and can do so automatically in a number of different situations, they can achieve productivity rates not achievable through more conventional methods.

IW team members simply fold to the empty slot and fill in as needed, or one or more members of the team will absorb the tasks of the missing person. It does not matter which person is missing from the team. When a person is missing, there is certainly going to be a loss of output, but there will not be a work stoppage. That is the most valuable attribute of Integrated Work Teams and is the main reason why they are much more productive than other types of teams.

IW teams that practice folding are also less susceptible to attrition than non-integrated teams because training is taking place at all times. IW teams receive training from one another. Strengths and skills learned only through experience are passed on from position to position in a natural way. IW teams are closer-knit groups because of their job sharing practices. They are less bored because they are not required to do the same task day after day. Integrated work teams usually have higher morale.

There are a few reasons that departments do not have integrated work teams. All of the reasons are understandable, but not necessarily acceptable. The most common reason is that companies quite often pay higher salaries to people that perform the more technical tasks that require more training. By doing so, they are perpetuating the existence of the specialist mentality. Companies do not want to pay equal pay to a technician and a newly hired person, and employees do not like to see a new hire receive the same pay as they are getting.

Proper compensation for IW teams can certainly create issues, but those issues are not insurmountable. Proper compensation can be achieved by adjusting pay scales according to the individual's background and experience. Employees that qualify as trainers could receive additional compensation. A company that encourages people to share their talents with others will be rewarded with higher performance. A company that encourages people to individually perform at higher levels will find that their employees will hide the secrets to their talents from others.

When a person is paid extra money to handle a skill position, it may not advantageous for them to help others not as skilled as they. Normally, when new people are hired, they are immediately put into positions that require the least amount of skill, or they are given some training and assigned to work with a more experienced employee. In a company that promotes competition among employees, they are not likely to receive quality training from their trainer. If they are fortunate enough to be assigned to an integrated work team, they will develop their talents much more quickly because it is in the best interests of the whole team to bring the new person up to speed as quickly as possible. If there is a Master Brewer in the IW team that they are assigned to, they are even more fortunate.

Even if there are no Master Brewers available, IW teams can absorb a new person and train them to handle all of the tasks much more quickly than a non-integrated team. The new person is trained in all of the positions because the team benefits when the new person develops quickly. As long as there is an unqualified person on the team that is unable to fold from one position to the next, the productive output is stifled.

Transforming a non-integrated team to a totally integrated work team is no simple task. The most difficult skills have to be mastered by every member of the team while still maintaining required production numbers. Training each team member to be qualified to handle the most skilled positions can be accelerated with Master Brewers if there are qualified candidates available. Once all of the tasks can be accomplished with a reasonably high level of skill, the team can rotate tasks periodically to reduce boredom and increase team skill.

Teams that are set up for Single Unit Flow and have the added strength of being fully integrated will experience output numbers that will be mind-boggling in comparison with the norm. Product will move through the work center like a sieve. Quality will improve and the effect of absent employees will be minimal. It is not uncommon to expect a 50 percent increase in output from this powerful combination.

Practicing The Art **And** Science of Manufacturing

Believing in all of these rules and concepts is one thing. Making them work for you is quite another. The business of production is not an easy one. If you had to pick a career, you could find a lot of possibilities that are a heck of a lot easier than manufacturing. Manufacturing people have to put up with conditions designed to insure their failure, yet they somehow prevail and actually find ways to get the job done. Quite often, they have people in critical positions that have no knowledge of manufacturing. When companies hire managers, they might look at the applicants' college credentials and experience, but they are not able to determine if the applicant really understands the science of manufacturing.

There is certainly a science to manufacturing, but because of the constantly changing conditions that exist daily, it is clear that there is an artistic side to it also. Finding the correct solution for a given problem every time requires a different kind of wisdom, but it can be done if our decisions originate from a sound foundational value set. The artistic part is to find a way to keep all of our decisions within that value set.

Understanding Little's Law and how to use it is extremely helpful, but sometimes problems crop up that make it very tempting to compromise our values for the easy quick fix. The solution may not be what we would like it to be but we feel that it is necessary to compromise at times. What we should realize is that we should always reject those knee jerk reactions and find that one solution that works **and** doesn't cause us to violate our principles.

Standing your ground under pressure is easier said than done. Managers have to make quick decisions – it is one of the requirements of the job. The better managers always seem to make the right decisions and the weaker managers struggle. When it comes to the world of production, the decisions could have a dramatic effect on the company's market share or its reputation. Finding the correct solution is much easier when we have a good value set to rely on. When we know beyond a doubt that we are doing the right thing.

We know that our solutions will work because we know that we are doing what is right, not what is convenient. Sacrificing our values may get us through the heat temporarily, but it will not deal with the source of the heat. Basing decisions on sound values that are consistent with good manufacturing practice is the proper course of action. Practicing the art and the science of manufacturing is certainly possible if we have that sound value set and we believe in it enough to never, ever compromise it. Couple that with a complete understanding of the science of manufacturing and you have a very powerful combination.

Chapter 3 – Process Mapping and Analysis

FCP has helped companies realize tremendous improvement, but it could not have done so without Process Mapping and Analysis. In each and every productivity improvement project, Process Mapping and Analysis, along with simulation modeling was an amazing catalyst for positive change. By presenting to management a strong business case for the implementation of changes that would improve the process dramatically, the changes were approved quickly. Without it, FCP could not have been successful!

How in the world can you possibly improve if you don't have a strong understanding of the details of your process? Don't you think that, in order to find your problems, you must first understand the details of what you do? There is no way, unless you are extremely lucky, that you can come up with solutions to problems and not have a detailed knowledge of your process. You must first understand the details of a process and have a way of evaluating every detail of it before you can determine what is right with that process, and what is wrong with it.

There are so many supervisors and managers that "think" that they know their processes, but do not. Supervisors might have a pretty good idea of most of the details, but managers hardly ever have a detailed knowledge of their processes. They only know what they think is supposed to be happening. And if they only have little more than a general knowledge of a process, how can they possibly be asked to improve their process performance? Well, they can't.

Managers and supervisors often live in a utopian world. No bad news from production translates into good news. It's one of the big reasons that they are so surprised and dismayed when something goes wrong. When there are no problems, they automatically assume that everything is running exactly as planned. They believe that everyone out there is following procedures exactly the way they should. While it is possible that everything **is** running according to plan, there is a strong possibility that the process could be slowly falling to pieces. They do not take the time to investigate in detail and thus are not dealing in reality.

An industrial engineer could do a study and determine that each machine is capable of producing 100 widgets per hour. Although consideration is given for a variety of conditions, the study may not reflect reality. The reason is simple: There are probably a multitude of time consuming tasks that slow production, or that too much consideration was given to conditions that do not exist. If flowcharting is used, and every single task is identified and documented, you have a different situation. Flowcharting combined with accurate numerical time data will result in representing quantitatively and qualitatively what **really** happens during the normal process.

If a process is in need of improvement, it is logical that the first step to improvement is to understand the details of that process. You cannot understand where you need to go until you understand where you are. Once you understand where you are, you must then determine what you need to change in order to get to where you want to be. All along the way, you use sound production values to determine what is right, what is wrong, and what the changes should be.

To understand the details of the production process, there must be a way of collecting, documenting and analyzing each step of that process down to the lowest level of detail. The collection of data is accomplished by interviewing the workers in detail. Each step of their duties is then documented in a way that facilitates analysis. Finally, each step of the process is subjected to a value add analysis that is based on good, sound production values.

There are a number of ways that a process can be documented. In the "old" days, they wrote procedures in a very technical way, using technical language that only the engineers could decipher. The technical procedures were placed into binders to be updated as necessary whenever a procedure was modified, or whenever someone was available to do the updates (if ever).

But procedures only covered the more complex or technical jobs on the line. Duties are added to the process, sometimes without the knowledge of management for reasons known only to the person who initiated the duties. The only time these "unofficial" duties were given any attention at all is when a new or untrained person needed to be trained or briefed on how and when these duties need to be completed. Sometimes technical leads and supervisors promulgate changes that become unofficially permanent because they were never rescinded. Promulgated changes are hardly ever documented even though they are often primarily responsible for productivity loss. Since they are seldom documented, they are not given much consideration in a productivity improvement effort.

Flowcharting is the perfect alternative to the written procedure. Written procedures are fine if you are into boring, heavy reading. First off, flowcharting is visual. It presents an image that is identifiable, which means that the intended message goes straight to the reader's brain – instantly. There is no confusion about the type of task being represented. Each task is documented with a symbol. Each symbol represents a type of process step and the details of that process step are contained within the symbol. Symbols send messages straight to the brain while written procedures usually use very technical language that is seldom understood completely and is therefore, often misinterpreted. See Figure 2.3.1 – Flowchart Shapes on page 124.

Collecting Data –
The First Step Towards Learning The Process

Before you can create a flowchart, you must first find a way of understanding the process and every single detail of it. You could rely on empirical data collected earlier, which gives the utopian or planned view of how the process works. You could do that if you were not interested in accuracy, that is. The only good way to collect the data is to interview the process experts. These are the people that do the work each and every day.

Collecting data through the interview is not as easy as it might appear. You go out there and ask questions, and the answers to those questions give you the information that you need. That is one scenario. The likely scenario would be that you go out there and ask questions and the answers that you get are the answers that the interviewees think that they should give. Obviously, there is a big difference between the two scenarios. Interviewees may hold back the truth, not because they are dishonest, but because they don't want to make waves with management.

In some plants you will find that there are as many "unofficial" process steps as there are official ones. A supervisor will temporarily initiate an order to add extra inspections, counts or redundant tests just because they want to make sure that mistakes aren't passed on to the next work center. There are many ways that process steps are added without the knowledge of management or process engineering. There are times that workers are asked to collect data for an engineer or supervisor and the person asking for the data completely forgets to tell the operator when the need for data has ended. The data continues to be collected even though it is not being collected and evaluated.

As noted earlier, there are tasks that take time away from production that were initiated by lead technicians, foremen and even group leaders without the knowledge of management. Some of these promulgated process steps are continued even after the initiator has moved on to another area or left the company. Nobody knows why they are performing these actions, and when asked how long they have been doing things that way, or why they have to do them at all, nobody has an answer. All they know is that it is required. Usually, the additional process steps add no value at all to the product and the process is better off without them. If, when collecting data, these unofficial process steps are not discovered, opportunity will be lost.

When machines, particularly test machines, are acting up and the operators can't get someone to fix the problems, the operators sometimes have to "manipulate" the machine to get it to work, almost like a youngster manipulates a pinball machine to get the ball to do what he or she wants. I've seen workers putting their whole weight on a test head because the machine didn't have enough downward pressure to make good contact. Should the worker be required to do that? No, of course not, but sometimes the operator feels that it is necessary in order to get their job done. They just cannot get maintenance or engineering support, so they take action on their own. In the case of test machines that determine the quality of the product, defects may be passed on to the customers. The main reason that operators try to work with the machine is that they want to avoid having to deal with false failures. What they don't realize is that while they are eliminating false failures, they might be causing the machine to miss real defects. Flowcharting will expose this problem.

People in responsible positions sometimes think that they are doing the right thing by telling their people to do redundant, or "just in case" tasks that take them away from their primary duties. Changes are sometimes made to the approved process during off shifts that are completely unknown to anyone except the people on the off shifts.

Many times these additional process steps are like hidden bottlenecks sitting in the background holding up production and nobody has any idea what the reason for the productivity loss is. Managers are at a loss to figure out why production has backed up because they do not know what everyone out there is doing. As far as management is concerned, everyone out there is following procedures. Well, they **are** following procedures – both the official and the unofficial procedures.

So how do you get the truth, the whole truth, and nothing but the truth? You go directly to the source of the truth. You only interview the people that actually do the work, and you conduct the interviews on their turf. You do not haul them into a conference room for an interview; you interview them right at their workstations where they will feel comfortable. When explaining what they do, they can actually demonstrate how they do their jobs. You do not want semantics and poor communication getting in the way of the truth.

The interview should be conducted in a way that is not threatening to the interviewee. You do not interview an employee while their supervisor or manager is nearby. The information that is provided can be tainted if the interviewee is nervous about his or her answers. Supervisors should never try to influence the interviewee before or after the interview in order to avoid being "discovered". Management should encourage the employee to be accurate and honest, nothing else.

In cases where production is pinched and the output from the work center is not meeting expectations, it always helps to make it clear that the goal of the mapping effort is to make things easier for the workers, not to make their work more complicated. The ultimate reason for flowcharting is to simplify the process and remove waste. If existing process steps are left out of the flowcharts that consume worker time, it will appear to some that the workers just are not as efficient as they should be, when in reality it is the process that is at fault. In many cases, the workers have so much extraneous work to do that production takes a back seat.

Starting at the beginning and finishing with the last process step, the interviewee should explain what they do step by step with every detail included. It is up to the interviewer to determine what is important to include in the flowchart after the data is collected. When the first cut of the flowchart is completed, the employee must verify that it accurately depicts what was given to the interviewer. If, after reviewing the flowchart, the employee requests changes, the changes should be made and again presented to the employee for verification. This process is repeated until the employee validates the flowchart as true and accurate.

Once the flowchart has been "blessed" by the employee(s), it is then important that the supervisor be shown the flowchart. The supervisor may not be happy about what they see on the flowchart, but if that is the way it is – that is just the way it is. Since the object of the whole process is to improve productivity, there is nothing wrong with making changes on the spot as long as the changes are useful.

In a process improvement effort, it would be dangerous to make a process mapping presentation to management that could not sustain challenges. Each flowchart must be absolutely accurate and representative of the process it depicts. When both the employee and the supervisor assign their signatures to the flowcharts, the probability of a challenge is minimal because the signatures attest to the validity of the flowcharts.

I was involved in a process-mapping project of a business where the company was paying thousands of dollars monthly for software, which was designed to automate the processing of home loans. When I presented my flowcharts to senior management, it was noticed that there were no data entry process steps in the flowchart. The vice-president whose budget paid for the automated system challenged my findings stating that there should have been multiple data entry points. He was paying big bucks for a system that his people were not using.

His challenge was not a good one, because as he later discovered, the automated system was not being used. What he didn't know was that there wasn't anyone that was knowledgeable when it came to using the system. As a result, the workers had reverted back to a manual process of filling out the forms – even to the point of using a typewriter as they did before the system was installed. Since this had been going on for some time, the cost was very high. The process mapping exposed the problem and it was fixed. The VP had no idea before the project that the expensive system that they were renting was not being used. He had been embarrassed but was not permanently damaged by the discovery.

The VP did not have to study the flowcharts, he could see by looking at the colors that there were no blue data entry operations on the charts and he knew that there were supposed to be a number of them. The discovery was instant, once he knew what to look for. A similar discovery could be made by looking at a flowchart and seeing too many red operations, which in FCP flowcharting depicts Non-Value Add. Seeing too much purple would indicate that there was too much transport. Shapes and colors make the flowcharts much more easy to read and understand. See Figure 2.3.3.

Shapes and Colors –
They Make The Flowchart Easy to Read

Examine each of the shapes used in FCP flowcharting. Each shape has a different meaning and use. You can tell what the nature of the task is by just looking at the shape. You know where the decision points are, and you can identify special tasks, such as paperwork, transport and data entry. The details of the task are written inside the shape and additional notes may be made beside the shape to add clarity.

GREEN

RED

BLUE

YELLOW

YELLOW

A

PURPLE

YELLOW

RED

RED

RED

RED

Standard operation - Green is the normal color for an operation that adds value to the product from a customer perspective (VA). The operation makes the product more valuable to the customer. Examples of Value Add operations are assembly, stamping, drilling, shaping and casting.

Standard operation - Red is the color for an operation that DOES NOT add value to the product from a customer perspective (NVA). Examples of NVA operations are filling out paperwork, inspecting, counting, and moving.

Data entry - Blue is the color for data entry into a system that is official data entry. Unofficial data entered into a personal database or computer is not covered by this symbol. Checking the status of an order on the system uses the same shape but is given a red border to indicate NVA.

Start/Terminate - The yellow border indicates that the shape is not a process step. The oval shape is only used at the beginning and the end of a flowchart.

Connector - This shape is used frequently to connect pages and process steps. The letter located within the connector shape shows the source of the last step. A is always matched to A, B to B, and so on.

Transport - This shape is used for the official transport of product from one area to another. It is only used when WIP is accumulated, then transported. The color for transport is purple. The reason for the unusual color is to bring visual attention to the amount of transport that is taking place.

Decision - This symbol is used to show that more than one path may be chosen and the choices are branched from the points of the Decision shape. Since the Decision shape is not an actual process step, it is also shown in Yellow. Examples of decisions are: Pass or Fail, and Inspect - Yes or No.

Paperwork - This symbol is used when there is official paperwork to be filled out that is in addition to the normal duties of building a product. The red border is used when the department is not in the business of producing paperwork as its final product.

Long Term Storage - When product is stored for what is considered "Long Term" periods, this symbol is used. Examples of Long Term Storage are Finished Goods areas and Filing cabinets where documents are stored.

Short Term Storage - This symbol is used for input and output queues where product is stored prior to the product being worked on or just after the product has been worked on. Other examples are short term delays, such as oven baking, waiting for adhesive to set, or product to dry after immersion.

Inspection - This symbol is used for Inspections and sometimes electronic test. When flowcharting a department whose primary responsibility is to test, this symbol would be shown in green.

Figure 2.3.1 – Flowchart Shapes

179

Color Coding for Value Add Analysis

The shapes in Figure 2.3.1 are color-coded. Color-coding is important in flowcharting because they make the flowchart much easier to read and understand. There are only five colors used in FCP flowcharting: Red, Green, Yellow, Purple and Blue. Each color-coded shape sends a message that adds clarity and understanding to the flowchart. As noted in Figure 2.3.1, some of the shapes can be either red or green. The green shape indicates that the process step is Value Add. It adds value to the product from a customer perspective. The red shape indicates that the process step is Non-Value Add. It does not add value to the product from a customer perspective. Remember that we are not talking about the external customer, but rather the internal customer.

When flowcharting, it is important to be able to differentiate between what is just **important** but NVA, what is **waste** and what is truly **value add** from an internal customer perspective. An important process step can still be non-value add. A process step could be NVA but be required by the company, the state or the nation. If you are mapping an inspection department, the inspection step would be considered value add. If you were mapping an assembly department, an inspection step would be considered NVA and the assembly steps would be VA.

The company for quality reasons might consider inspection necessary, but quality has a cost, and the cost of inspection is added to the price of the product. In a tough competitive world, adding a nickel here and a nickel there could make the product cost too high. The extra NVA step also increases cycle time. This wouldn't seem so bad if there were only a few NVA steps to worry about, but that is not normally the case. With competitors looking for an edge, cutting out the waste is what could make a company tough to compete against.

Having said that, for the company that has not found a way to build flawless product without test or inspection, those extra steps are necessary evils. Remember, the best way to eliminate process steps and improve the process is to eliminate the **need** for the waste. You cannot just eliminate a process step regardless of what it is without investigating the consequences of the removal. The step(s) that are targeted for elimination may just be essential to the quality of the product **right now!** Find a way to eliminate the **need** for the steps and then eliminate them.

Inspection and testing is necessary in some companies because they have not found a way to build product without occasional errors. With some product, it is absolutely critical to the survival of the company to make sure that error free product is shipped to the end customer. In order to catch the mistakes early in the build process where it will be less costly to fix, inspection and/or test departments are added to the line. When mapping a test or inspection department remember that their end product is inspection or test. In test or inspection departments, the symbol for test or inspection would be green because testing or inspection adds value to the product for that department.

Redundant anything is certainly Non-Value Add, and is probably Waste. **If** testing the product twice just to "make sure" that the product is good is sound logic, then testing it 3 times is even better logic. Testing it four times would then be twice as good as testing it twice. The notion may seem crazy to some, but such logic not only exists, it is strongly defended by some. They argue that "their " product is so complex that multiple tests and/or inspections are necessary. They are wrong in their logic and they are wrong in their assumption that the complexity of the product is the problem.

The problem is in the inspection or test process. Improve the test and it catches the problem. Improve the inspection and the defects are caught the first time – not on the second or third try. For those reasons, redundant anything should be mapped with red symbols. If, during the analysis redundant process steps are tagged as non-value add, they will be targeted for extinction and hopefully eliminated by eliminating the need for the redundant steps.

The first step in value add analysis is to determine what the **product** of the department is. What does that department contribute to the company and its product? If the department is an assembly department, their product is assembling. If the department is a paint department, painting is their product. On the other hand, if your department is the test department, your value add steps are testing. A drilling department has a responsibility to drill and all of the steps where drilling is conducted is value add. The drilling steps are value add (Green) and the preparation steps are non-value add.

Setups are NVA. The customer will not pay you more for preparing the drills for drilling. Neither will they pay you more if you inspect the product to make sure that all of the required holes were drilled. If you want to add more inspectors, the end customer will not pay you more for your product. All that they care about is that the product be defect free. How you accomplish that is your business.

In flowcharting, be very critical about what you consider value add. Red presents opportunity for improvement through the removal of NVA so therefore you want to be able to identify the non-value add and have the ability to exploit the opportunity. But not all non-value add steps are created equally. Non-value add steps fall into two categories, each of which present a different level of opportunity.

The two categories of Non-Value Add are:

Non-Value Add Essential (NVA-E)
Non-Value Add Non-Essential (NVA-NE)

Non-Value Add Essential is another way of saying that the customer won't pay more if that process step were completed, but the government or the company requires that it be completed. Examples of NVA-E steps are adding UL stickers to the product, inspection tags, and quality stamps placed on the product. All of these do not make the product more valuable to the customer, but manufacturing is certainly required to complete the steps or the product cannot be shipped.

Identifying a step as NVA – NE is the same as saying that the company would be much better off if that process step would go away. Non-Value Add Non-Essential presents the greatest opportunity because any process step that is tagged as NVA-NE is probably **Waste** and should be eliminated. But knowing that it is waste does not mean that you can just go out there and eliminate that step. First you must find out why the step exists and then you find a way to eliminate the need for that step. It may turn out that you can just eliminate the step because it is so wasteful that the instant removal is the most logical solution.

The determination between NVA-E and NVA-NE is not made during the initial flowcharting process. At that time, the only determination made is that the process step is NVA and is therefore shown with a red border. Later, when the steps are identified as NVA-

NE, there are several options, such as writing the NVA-NE above the shapes or making the red border thicker than normal. In each case you would be making a statement that the process steps present the most opportunity to process improvement with their extinction.

Color-coded flowcharting is an effective way of documenting a process because the colors make the chart easier to understand and evaluate. If there is a lot of red, there is opportunity. If there are a lot of purple transport shapes, there is opportunity for improvement. Black and white flowcharts do not provide the added dimension of instant understanding.

Figure 2.3.2 shows what a flowchart looks like before value add analysis and Figure 2.3.3 shows a flowchart after value add analysis. Notice how color-coding can make a flowchart easy to follow, and show opportunity.

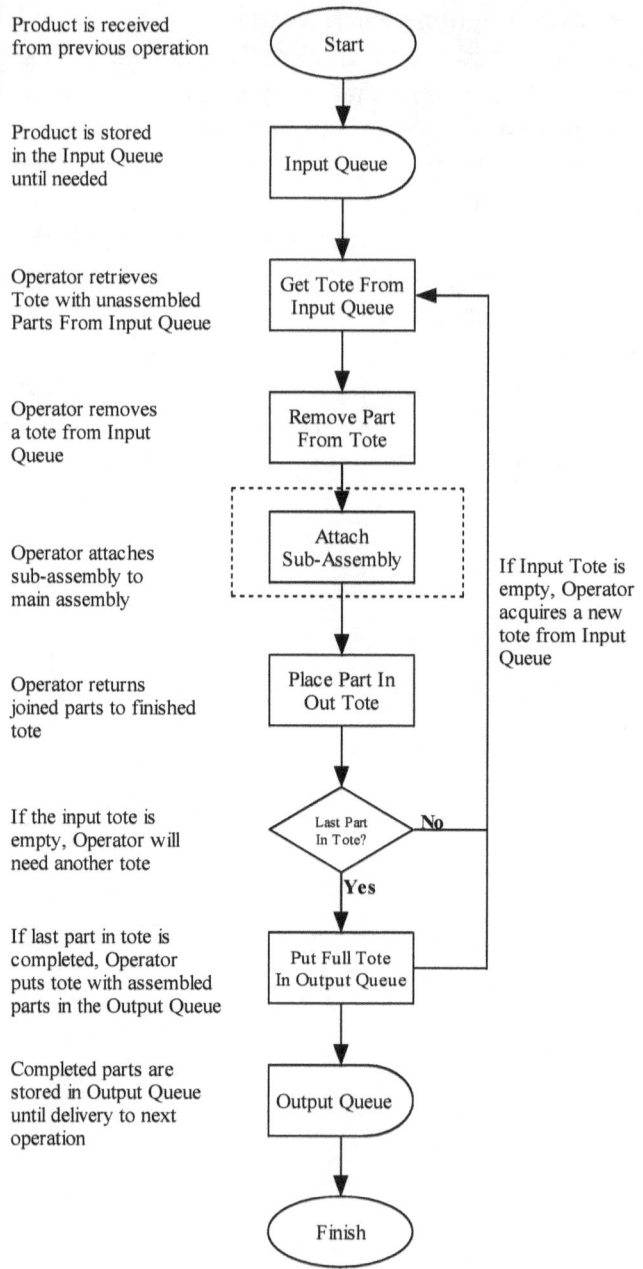

Product is received
from previous operation

Product is stored
in the Input Queue
until needed

Operator retrieves
Tote with unassembled
Parts From Input Queue

Operator removes
a tote from Input
Queue

Operator attaches
sub-assembly to
main assembly

Operator returns
joined parts to finished
tote

If the input tote is
empty, Operator will
need another tote

If last part in tote is
completed, Operator
puts tote with assembled
parts in the Output Queue

Completed parts are
stored in Output Queue
until delivery to next
operation

If Input Tote is
empty, Operator
acquires a new
tote from Input
Queue

Example of a Simple Flow Chart
Figure 2.3.2 – A Flowchart Prior to Color Tagging

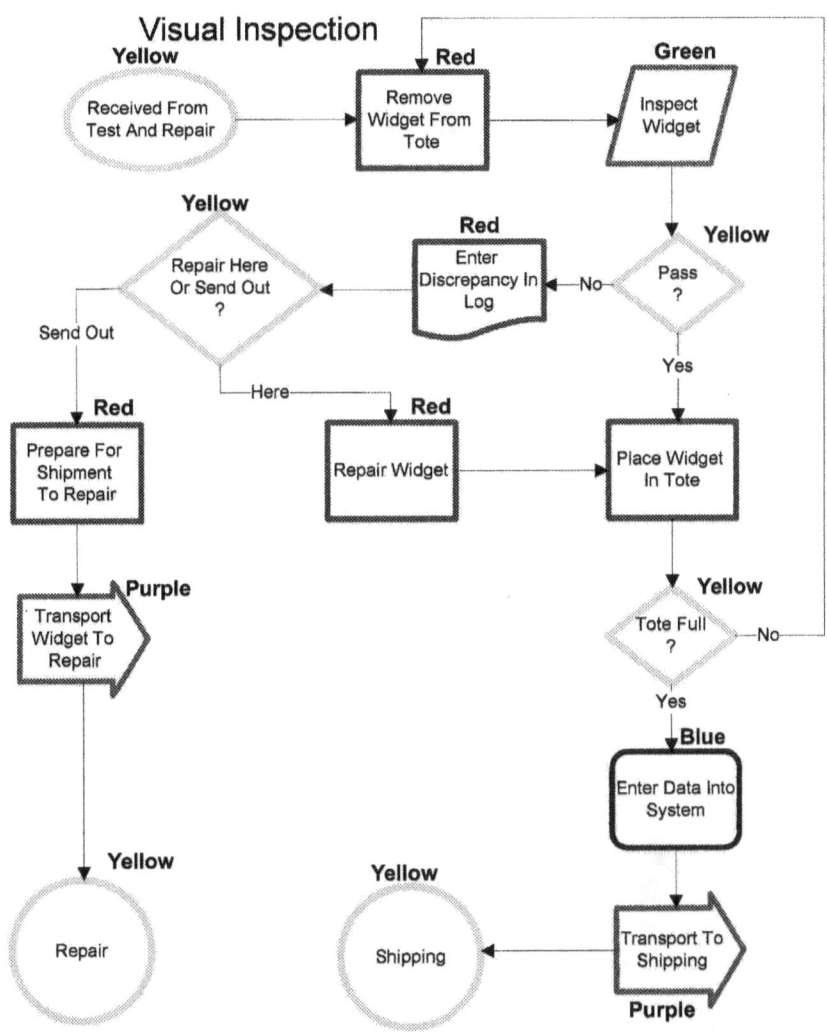

Example of a Flowchart After Value Analysis
Figure 2.3.3- Color Coded Flowchart

Tom Clason

Part III – TAKT Management – The Best Continuous Improvement Tool Ever!

Tom Clason

Chapter 1 – Elements of TAKT Management

Becoming a world-class producer is no easy task, but it certainly is rewarding. Your competitors wish that you would find another line of work, and they worry for their future. But continuing to be a world-class producer is even harder. In order for that to happen, you need to have to have a continuous improvement process in place. You need to have a way of finding problems as they crop up, and if you want to stay competitive long term, you must have an effective way of dealing with problems.

TAKT Management (TM) can both help you become a world-class producer, and help keep you there. It is, in my opinion, the most powerful continuous improvement tool that exists today. At the beginning of an improvement project, it is one of the most important tools that can be introduced to the process. Once the project has met its goals and the company has migrated to a day-to-day business environment, TM will continue to be the most important tool that you have. TM will help you to improve your operation each and every day … provided you continue to use it, and use it properly.

As I said, getting there is one thing, staying there is quite another. The successes that you achieve after the big gains will be smaller, but no less significant. During the cycle time improvement project the huge obstacles are overcome and there is a tendency to think that the war is over once the project is completed. That kind of thinking could very well result in a reversal of most, if not all, of the gains achieved during the project. Instead of continuing to improve, you will find yourself in an old familiar position: unproductive and in trouble.

Companies everywhere look for something that will get them to the next level. They seek that one little thing that is going to turn everything around for them. They don't mind working at it, but they definitely want results. Well, here's some bad and good news: There isn't **one** little thing that can get them to the next level, but there is one tool that can surely help them to improve their chances. TAKT Management can help their process to be more disciplined and it gives them a way of identifying the strengths and weaknesses in their processes. Takt Management is only one of the FCP tools, but it is certainly one of the more important ones.

The Beat!

Takt, simply defined is a German word that means, "Beat", as in drumbeat or heartbeat. How fast is the beat? In music you dance faster when the beat is fast. Your heart pumps more blood when the beat increases and your production line puts out more widgets when your Takt (beat) increases. Knowing your Takt tells you something, but by itself does not tell you everything you need to know about your operation. It doesn't solve any of your problems, but it tells you that you have a problem; it lets you know where it is, and why you are having it. It even tells you how much it is hurting you. The only thing it doesn't tell you is how to fix the problem. You need to rely on other tools and your expertise for that. Takt can also tell you if something is going well and why. It won't ensure that you will continue to do well, but it tells you what caused you to do well. Knowing your Takt is the first step towards continuous improvement, but there is more to it – much more. Takt, by itself does not tell you how to solve your problem, but it can quantify the effect that the problem is having on your production, and this becomes the first important step.

Takt – As In Output

Before you can determine your beat, you must have one! The beat of your line is the movement of product from department to department. Many lines have erratic product movement. Movement of product may take place one hour, for example, and not again for several hours. The beat is irregular to say the least. An unpredictable or irregular beat is like an irregular heart beat – you don't want it to be yours. Predictable is good, and in production, you want it to be predictable and substantial. You want your beat to be sufficient to meet the demand. To have a beat in production, you must have a constant movement of product from department to department. Hopefully, the amount of product being moved will be sufficient enough to meet your demands. If the beat of your line has not met the demand for your products, then at least you know it, and with TAKT Management you can do something about it.

The TAKT Chart

Takt is measured in units of output per hour. That is, units per hour **that were shipped to your customer.** You can look at the output of the entire factory, but it is much more effective to use TAKT Management across multiple departments within your factory. To illustrate how TAKT Management works let's assume we are a specific department in an imaginary widget factory. You may have completed 100 widgets, but if only 30 were given to your customer that hour, your output for that hour is only 30. The next hour you might ship 170 and your average, or your Takt for the two hours, goes to 100. Hour 3, 4, 5, 6, 7 and 8 yield 80, 120, 100, 50, 150 and 100 respectively. Let's look at the hourly output for that shift:

Hour 1 - 30
Hour 2 - 170
Hour 3 - 80
Hour 4 - 120
Hour 5 - 100
Hour 6 - 50
Hour 7 - 150
Hour 8 – 100

The total output for the shift is 800. When we divide that by the 8 hours in the shift, we get 100. (By now you've probably noticed that I like it when the math is easy) The Takt (average) for Department A is 100. Now let's look at a chart of the Takt for Department A.

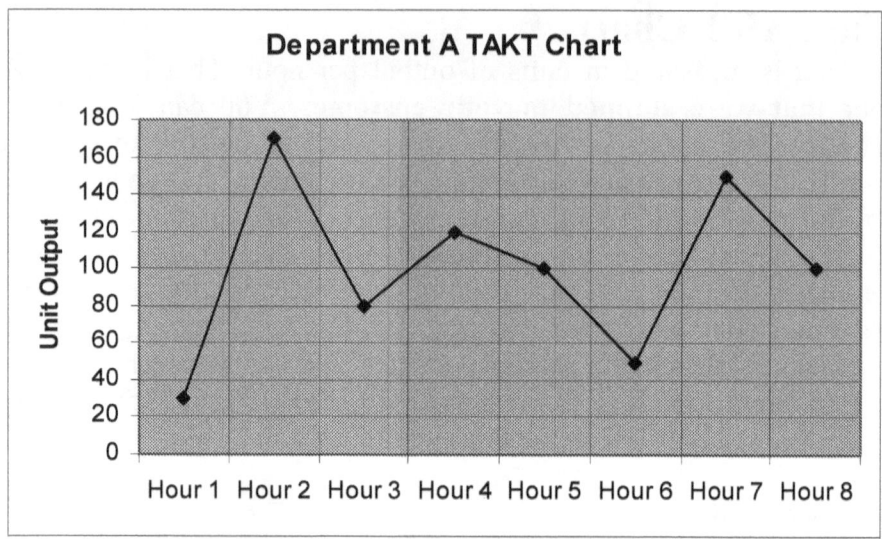

Figure 3.1.1.

Although I'm calling Figure 3.1.1 a TAKT Chart, it really isn't one ... yet. Yes, we know what our average output was for that shift, but we don't know enough yet. We only have that one variable – output. We don't really know what happened out on the line yet. Looking at the chart, we see some peaks and valleys. On hour 1, only 30 units were shipped followed by 170 units in hour 2. The same situation shows up in hours 6 and 7 where 50 and 150 units were shipped.

What could have happened in hour 1? Perhaps only 30 units were shipped because some of the product was held up because the person responsible for making the delivery was on break and waited until hour 2 to deliver the balance of the product. This delivery scheme could have created some problems for Department B, their customer. What if 30 units were not enough to keep Department B busy? They would have run out of product to work with, thus falling behind on their daily output. Sending them 170 units the next hour might not help them to catch up because they may not have the capability of catching up. They may be the bottleneck!

Now, this situation could be a recurring problem that happens every single day. Every day, Department B loses production output because Department A just doesn't give them what they need, when they need it. We wouldn't know how much of a problem that it is unless we were able to determine what happened with the peaks and valleys. Department B's TAKT Chart would show that they lost output because they ran out of work.

Unless the output for a given hour is 20% or more off the average, there is really no reason to give a "reason" for the delta from average. I have used 20% as a guideline. The proper percentage may be higher or lower, but when the delta exceeds the accepted upper or lower limits, it is absolutely essential in TAKT Management that there be a reason given. We refer to these reasons as Enhancers and Detractors, which I will further describe in a moment. But even these reasons are not enough – there has to be a way of archiving the data so that it can be evaluated later. Over a period of time, it will be possible to retrieve the data and evaluate it. From the evaluation, solutions to the problems will be found resulting in process improvement. It is the ongoing use of these TAKT Management tools that enable a process of continuous improvement.

As I've stated previously, the TAKT Chart (Figure 3.1.1.) is incomplete. In order to be a true TAKT Chart, it needs some additional information. It needs upper and lower limits to help identify output anomalies as well as explanations for these peaks and valleys. We will call our problems (valleys) "Detractors" and refer to positive events (peaks) as "Enhancers". Whenever the output is above or below 20% of the norm, the reason for the Enhancer or Detractor must be recorded on the TAKT chart. In our example, the Takt average is at 100, therefore any output amount below 80 or above 120 requires an explanation. Figure 3.1.2 shows how the Enhancers and Detractors are placed on the chart:

HOUR	REASON
1	Resources - Delivery person on break
2	Resources - Left over product delivered
3	
4	
5	
6	Quality - Product failed inspection Rework required
7	Quality - Rework completed
8	

Figure 3.1.2 – Detractors and Enhancers

Now let's look at Figure 3.1.3 and we can see a much more complete TAKT Chart. The Takt average (Red), and the Upper Limit and Lower Limits (Blue) are shown on the line chart, as well as the two elements: Hourly output, and Detractors/Enhancers (lower chart).

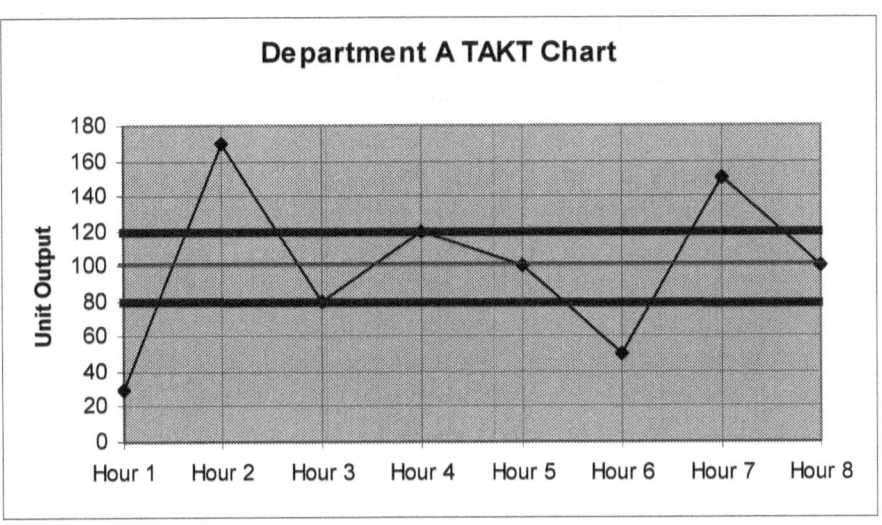

HOUR	REASON
1	Resources - Delivery person on break
2	Resources – Left over product delivered
3	
4	
5	
6	Quality - Product failed inspection Rework required
7	Quality - Rework completed
8	

Figure 3.1.3

Now we have something that we can work with. We have our hourly output, average output and the reasons for any aberrations in the output. Notice that the peaks and valleys in the output part of the chart are explained on the lower section of the chart. Although this charting technique is great stuff, which allows us to look at our progress on an hourly basis, it still is not enough. We still do not have the foundational element of TAKT Management that makes it so powerful: database archiving. We only have a daily TAKT Chart.

The chart itself is still incomplete. We need more information on the chart. We need things like the shift and the date. Some companies show the previous day or previous week's average TAKT on the chart, which is ok. It is not a good idea to put the "required" or "needed" amount on the chart. I consider that another way of pushing for numbers, which is a clear contradiction of Deming's principles. Pushing for numbers is not what TM is all about; it is about creating an environment of continuous improvement. Anytime a production "bogey" or numerical quota enters the TAKT Management process, the continuous improvement process quickly exits. Achieving the numerical goal, at any cost, becomes the prime objective. The prime objective in TAKT Management is to continually improve the process.

Using The TAKT Chart To Promote Positive Change

TAKT Chart designs differ from company to company. One company prefers three 8-hour charts for their 24 hour, 3-shift situation, and another company prefers two 12-hour charts. For a company with three shifts, I personally prefer the 12-hour TAKT Chart because it helps to eliminate competition between shifts. The emphasis should be to find and fix problems, not worry about which shift is doing well, and which shift is struggling. It will become clear enough to all if one shift is either very productive or having problems. There could be a variety of reasons for higher or lower output on each shift. The staffing may be unequal or the maintenance and engineering support may be lacking on the off-shifts. TAKT Management can help identify the need for these resources and bring the right focus towards addressing those needs.

There are undoubtedly problems that happen consistently on certain shifts, which TAKT Charts will help to expose. Excessive machine downtime that is associated with a lack of maintenance support on the off shifts is an excellent example of shift related problems. A lack of engineering support on off shifts or the seizure of machines by engineering on the day shift because engineers do not work the off shifts. Quite often, the maintenance or engineering seizure of machines on first shift is the very reason that output is higher on the off shifts, particularly the second shift. When these kinds of problems are recorded on the TAKT Charts and archived, finding positive solutions will soon follow.

A very common and often ignored Detractor is shift change. There is always a very significant output loss when shift change takes place. At some companies, the output loss starts as much as two hours before the shift change and continues two hours after the change. There are a variety of reasons for the loss of output due to shift change but the one that really stands out in my mind is associated with machine setups. I experienced this problem with several companies. In each case, the problem was a significant one that was solved through TAKT Management. The TAKT Chart for this problem always showed that there was very little output one hour before and one hour after shift change. The Detractor entered on the TAKT Chart showed, "Shift Change". The output for the two hours surrounding the shift change was 50% or more below average.

After an investigation of the Detractor, it was found that the following conditions existed:

- The incumbent shift would not start a new order within one hour of the end of their shift.
- If an order started by the previous shift was not completed by the end of the shift, the relieving shift would set the order aside.
- If an order had quality problems related to a setup, the person that finished the order would be responsible for the quality of the product, not necessarily the person that performed the setup.
- The setup procedures were not clearly documented and enforced. Each employee had his or her own way of executing the setup
- The shift change itself was not documented or well defined.

When management responds to quality issues by writing up the shift that completed an order, it is possible that the wrong person may get written up for a quality problem. Once the wrong person is written up, workers will be reluctant to work on orders that were started by workers on other shifts. The problem could get so bad that the shift going off duty will refuse to start new orders within an hour (or more) of shift change unless they can complete the order themselves. There is no reason to start a new order if it will remain untouched until your shift returns to work.

It always amazes me to see just how much output was lost at those plants because of shift change, but even when the problem was solved, it was even more amazing how difficult it was to implement the solution. The TAKT data quantified the effect of the output loss. Everybody there knew that the problem existed, but Manufacturing and Quality management could not agree on a way to deal with the quality issues associated with faulty setups. The root cause was identified as inconsistent setup procedures that varied not only from shift to shift, but also from person to person. Everybody had a different way of doing a setup. In other words, the people doing the punishing were the ones that should have been punished. Management and engineering simply did not have clear and up to date procedures for doing setups. The setup problem was eventually fixed through FCP Team Setup reduction (See Part 5 - Setup Reduction – The Team Method of Setup Reduction), but the problem itself was initially identified through TAKT Management.

There are solutions to these and other problems, but there has to be a way of determining if, in reality, they really are problems. The best way to understand the magnitude of the problem is to quantify the occurrences and the productive loss associated with those occurrences. Remember, a Detractor doesn't appear on the chart unless the output is 20% or more below the norm. If a Detractor appeared every single day, the output loss for that Detractor is at least 20% of the total output. If you can understand exactly how much the problem cost in terms of output loss, you can then determine which problems are serious, and which problems are temporary aberrations. TAKT Management helps you to understand the magnitude of your problems and fix them. When you start eliminating the output loss, you are improving your productivity and your profit.

Little problems, when constantly repeated, become big problems, but often are not seen as such without TAKT Management. They don't get the local press that the really big problems get because nobody understands the true impact that they have on output, quality or profit. Until a problem gets serious attention, they are pretty much ignored. Everybody loves to dive into the big issues, because once they are solved, somebody is going to be a hero and high fives abound! But fix what is considered a little problem and everyone just ignores it. With TAKT Management, a company discovers which little problems are really big problems. TAKT Management helps to understand just how often the so-called "little" problem happens and the net output loss when it happens. For example, if just one hour of output is lost every single day, then around a full day of output is lost each month. How many companies feel that they can afford to lose a full days production needlessly?

Floor personnel enter all of the TM data onto their charts; the accuracy of their information becomes critical. There are times when they are either not well trained or do not know how to make a particular entry. They may be filling in for someone that has not trained them properly, or do not appreciate the importance of the charts. The best way to avoid these types of problems is through comprehensive, across the board training. **Everyone** should be trained in the details of TM and how to make entries on TAKT Charts. When people are properly trained, you can be sure that the accuracy of the charts and the effectiveness of TAKT Management meetings will improve.

It is absolutely essential for the charts to reflect the truth, but in some cases, a false entry may be made on purpose. Even though people make mistakes that cause output loss, you may not see the words "Operator Error" on a TAKT Chart unless such an entry is not seen as threatening by the operators. If the operator or a supervisor feels that such an entry would hurt their careers, they may be highly motivated to enter another reason for the output loss on the TAKT Charts. If the entries on the charts are not accurate, the accumulated data could lead to faulty conclusions later.

This is where management can help by understanding the situation and making it very clear to the employees that honesty will result in help, not reprisal. An occasional operator error is a fact of life and management should be looking for ways to make errors impossible

199

through positive action such as training and supervisory involvement, not through the fear of being caught. You can be sure of one thing: If employees are afraid of being punished for something that they put on the chart, even if it is the truth, They will be reluctant to put themselves at risk. On the bright side, management can avert this potential problem by attacking the problem, not the people.

People Are <u>Not</u> The Problem!

Here is an example of two supervisors with different companies who reacted in very different ways to the entry of "**Operator Error**" on a TAKT Chart. The first supervisor was most interested in "who" made the error. When he found out who made the mistake, he called the guilty person into his office and read him the proverbial riot act. He made it clear that the mistake was "unacceptable" and that it had better not happen again. "Operator Error" never appeared again.

I am convinced that the problem did, in fact, happen again, but it never did appear on the chart. The problem was not solved; it was merely covered up. That supervisor had no idea how much damage he had done. I found out about the incident several months later through one of the employees and had a private conversation with the supervisor. Just like the employees that worked for him, the supervisor was really a good person trying to do what he thought was right. He said that he would try to correct the situation, but with these employees the damage was done, the individual employee's trust was lost, and the word was out; causing the entire department to be on guard. That supervisor would have trouble gaining trust with his employees in the future.

The second supervisor handled the problem quite differently. This supervisor called all of the workers to a meeting where he asked one question: "What can I do to help make this problem go away?" He made it clear that he was not interested in finding out exactly who the guilty person was; he was interested in correcting the situation. The employees were open and honest and gave the supervisor good feedback because they trusted him and didn't fear reprisal from him. The supervisor took the advice of the workers and the problem was corrected. The workers felt much better about their supervisor and much better about filling out their charts honestly. They weren't intimidated to avoid mistakes; they improved their abilities because they wanted to. They were properly motivated to improve and their attitude towards quality was improved. Unlike the first example, the improvement would last.

One of the basic rules for filling out a TAKT Chart is to never put a person's name on the chart. Putting someone's name on a chart will damage the usefulness of TM faster than just about anything. Even when you do not put a name on the chart and a keyword like Operator Error appears, it may not be hard to figure out who made the error. Workers do not like to be identified as the creators of problems and they will sometimes go to great lengths to avoid that kind of exposure. Sometimes the employee who made the error is not the one that is entering information onto the TAKT Chart. The person that maintains the chart may hesitate to "rat" on his or her co-worker making these situations very awkward and even stressful for the employees. If management creates the proper atmosphere, this kind of problem simply will not exist.

Even the most diligent and careful employee can make mistakes on occasion, and the most likely cause of the error is directly related to poor training or lack of experience. It is very rare that a problem is directly related to blatant incompetence or indifference. It is so easy to say that people should be encouraged to help solve the problems, not work to cover up their mistakes, but it is more difficult to provide training that will convince them of the importance of accuracy. The difficult path should always be taken because it will pay the highest dividend. TAKT Charts should never be a vehicle for finding guilty people because it will destroy the effectiveness of the tool.

The TAKT Database –
The True Power of TAKT Management

The most useful element of TM by far is the recording of all hourly outputs and all Detractors and Enhancers into a central database. But data in and of itself is insufficient. Data must be converted into useful information in order for the business to analyze it and formulate improvement plans as a result of its use. The TAKT database can be an effective tool in terms of gathering data and outputting useful information for analysis. By entering the data into a database, you can retrieve the data in meaningful ways at a later date. Modern database programs are easy to use and allow for the creation of special reports. Here are some examples of the kinds of reports that can be created:

- Average hourly output (24 hour report) – The recording of the average output for every hour of the day. You can understand the effect of shift changes, lunch and even meetings on output.
- Average daily output (Monday through Sunday) – An excellent report for management to understand the effect of weekend overtime and startups.
- Top 5 Detractors and Enhancers – It is important to understand what is working well and what problems just will not go away. When a detractor makes the Top 5 list for a sustained period, this report will bring attention to it at TAKT Meetings.
- Average output for all of the occurrences of specified Detractors and Enhancers.
- Machine Setup Times can also be analyzed, provided setup data are recorded as a Detractor for a machine operation.

These and other reports can be created that provide meaningful insight into the many problems that exist in production and the frequency of their reoccurrences. Understanding the net effect of the reoccurrences is the first step but it is a very important first step. You have discovered where the hurt is; now it is time to make the hurt go away. It is time to take the information that has been recorded, determine what problems are the most serious and then fixing them. TAKT Meetings is the next step.

TAKT Meetings – The Instruments of Positive Change

The solution finding elements of TAKT Management are the daily, weekly and monthly TAKT Meetings. Each of these meetings serves a very defined purpose. The daily meeting provides a snapshot of daily issues on the production floor and allows flash problems to be identified and dealt with in a timely manner. In addition, the daily meeting makes it possible to deal with daily issues quickly before they become bigger problems. The weekly meeting is used by management to attack problems that cannot be solved at the daily meeting. The Weekly TAKT Meeting is also a good way to analyze weekly data summaries. These weekly summaries begin to identify patterns or common themes, which are not evident in the daily meeting, and a stronger sense for the impact of problems, are also realized.

The monthly meeting is used primarily by executive management to both monitor the results of the weekly meetings and intervene into the solution of problems that simply will not go away. The monthly meeting keeps senior managers aware of the problems and successes experienced during the month, thus allowing them to be part of the TAKT Management solution. Here are some basic rules for each meeting:

Rules For The Daily TAKT Meeting

Here are some basic rules for Daily TAKT Meetings:

1. There should be a room dedicated to nothing but TAKT Meetings.
2. Training is essential! Each presenter should receive training in TAKT Chart presentation.
3. Presenters should be selected carefully. Each presenter should have credibility, have good presentation skills and be willing and able to make presentations to an audience.
4. Do not select presenters that will cause the areas that they come from to suffer a loss of productivity with their absence.
5. Presenters should prepare themselves well for their presentations. They should be prepared to answer pertinent questions about the entries on their charts.
6. Attendance for representation from each support group including management should be mandatory.
7. Meetings should be held to approximately 30 to 60 minutes.
8. Whenever questions cannot be answered on the spot, action items should be given to provide answers within one working day. Action items should be resolved within the current workweek.

9. The meeting is not a forum for debate. The meeting should have strict rules and a moderator to take issues off-line if they cannot be resolved quickly.

10. Names are never associated with Detractors at the meeting. When a person is responsible for an Enhancer, or for a problem being solved, that person should be recognized and applauded for their involvement. Recognitions should take place on the spot, and in a timely manner. Celebration is encouraged at all meetings.

The first thing that is needed is a room dedicated to nothing but TAKT Meetings. Unlike Weekly and Monthly meetings, the room has to be designed just for the purpose of TAKT Meetings. The room should be large enough to accommodate each area's TAKT charts that will be presented, so that the audience can see each of the charts. The charts should be laid out in the same order that each work center has in production. Prior to the start of the meeting, each person attending the meeting should arrive early enough to get an advanced look at the charts. If any of the charts raises a question in their mind, they can bring up the question either before the meeting, if possible, or during the meeting. If an entry brings a question to the mind of one person in the audience, it will probably bring up a question in the minds of others at the meeting.

The daily meeting may really be a multi-daily TAKT Meeting depending upon the plant, its size, and the number of departments. It is important that each shift participates in TAKT Management, so some companies may opt for three meetings daily. Other companies may decide on two meetings, one held between the night shift and the day shift to ensure participation by the night shift, and the other held between the first and second shifts. Companies with only one or two shifts may opt for either one or two meetings. Either is fine as long as each shift participates in the meeting. A company with two shifts should either have one meeting between shifts or two meetings – one at the beginning of first shift, and the other at the beginning of second shift.

The Daily TAKT Meeting starts with a presentation of the TAKT Charts. The presentation is normally given by non-management personnel – usually leads, foremen and the like. Some companies prefer to have their supervisors give the presentation because they want their production people on the floor. Either is fine, as long as production is not being adversely affected. These presenters may be giving a presentation for data collected on another shift, so there has to be some coordination to ensure that the presenter is able to answer questions on Detractors and Enhancers that occurred on another shift.

The meetings should not be a forum for confrontations, but should be a time where problems are solved and issues are discussed. The daily meeting is an in-depth look at the output and issues for the past 8 to 12 hours. It is a time when people get together to study the issues and learn from them. The meetings should be less than an hour – preferably 30 minutes if possible. Long meetings are never as effective as short meetings as long as the short meeting is able to accomplish its objectives. To keep meetings as short as possible, there must be a facilitator with an agenda to keep the meeting moving. As noted, each presenter should be able to answer questions about the charts and explain the details of each Detractor or Enhancer. At times the Detractors and Enhancers that are presented on the charts may be vague or misleading. The presenters must prepare themselves in advance so that they understand their charts before making the presentation not just show up and read what they see.

When a question or a challenge to a Detractor or Enhancer is raised, the issue should not become a debate during the meeting. The TAKT Meeting should solve issues, not create new problems. The best way to deal with time consuming issues is to hand out action items to someone in attendance to investigate the issue and report back at the next meeting. If a Detractor identifies a problem that can be dealt with quickly, one or several people at the meeting should be given the responsibility of looking into it. As noted, this is a meeting where problems are identified and solved as quickly as possible.

The TAKT Meeting should not be a forum for sales or production control people to lobby for action related to customer issues. This type of intrusion is harmful to the meeting because it shifts the focus away from eliminating obstacles and learning from successes. I have seen the production control personnel successfully request that certain orders be prioritized. In one plant, the production control organization actually dominated the TAKT Meeting spending more time on their issues than on the presentations of the charts. TAKT Management eventually took a back seat to prioritizing at this plant, which eventually destroyed the original intent of the meeting.

For TM to be an effective tool, the purity of the purpose must be maintained. As powerful as this tool is, it is also vulnerable to corrosion and corruption. When Production Control takes over the meeting, it will destroy TM as a continuous improvement tool and eventually render it useless. Another company tried to include production control in the meetings in a small way. Initially, production control, started out with 5 minutes to present their issues. Their presentation time kept increasing until they were taking 30 minutes and sometimes more. They were taking so long talking about order issues that there was no time for anything else. The charts were nothing more than symbolic pieces of paper that nobody looked at. All that was addressed was a list of orders that needed to be pushed, and what needed to be set aside. Instead of improvement, what they accomplished was productive gridlock. The destruction of TAKT Management was complete at that plant.

Although it may seem that I am trying to exclude Sales and Production Control from the Daily TAKT Meeting, the opposite is really true. Their involvement in TAKT meetings is very important. They simply should never control or dominate the meetings. An order to prioritize should never originate at a TAKT Meeting at the request of a support group. Each support group representative should be there to help promote proper production values and not ask that action be taken that is in violation of those values.

Depending upon space limitations, the audience should consist of one or more representatives of each and every support group including Maintenance, Engineering, Production Control, Sales and Management. The business of production is a team effort and all of the teams should be represented. Upper management should also pop in periodically to show their support for the effort. One company would occasionally bring their customers to the meeting whenever they were in the plant. The presence of customers and senior management at the meetings was gratifying to the in-plant audience and motivated them to improve even more. It was clear to all that the visitors were quite impressed.

Not only should support representatives attend the meetings, they should participate by taking action items to issues that pertain to them. They shouldn't just sit there and say nothing, but should be an active participant in the meeting. They should always be openly supportive to the improvement effort. As I said, everyone that works in the factory benefits when production improves. It is much more enjoyable to work in a successful factory than one that is wrought with serious problems. Failure is a word that none of us care to be associated with. We all want to be successful and we all want to participate in success. TM allows us to do just that. Working to improve the process is far superior to addressing the status of orders. If you work to improve the process, everything else falls into collateral benefits.

Weekly TAKT Meetings

As powerful as they are, daily meetings cannot solve all of the problems that crop up on a daily basis. Some issues require involvement from middle management. But management should be working on meaningful issues and problems that just won't go away. Daily meetings will not solve a problem that crops up only now and then. For those problems that require investigation and analysis to solve, there is the weekly meeting.

The weekly meeting is a perfect way to address lingering issues. Detractors and Enhancers can be Pareto analyzed to help the team prioritize its improvement efforts. These reports are issued weekly and each supervisor is required to address the top 3 or 5 Detractors and Enhancers for their area. Each week they report, one by one on their progress. The supervisors report their findings and answer questions from the middle managers. Each week, the supervisors are working to eliminate the problems that inhibit production. This is the beginning of continuous improvement.

By dealing with only the more serious Detractors (Top 5), the supervisors are focusing their efforts and solving problems one by one. While it is true that not all issues can be solved at weekly meetings, it is the ones that cannot be solved at this level that are brought to the attention of senior managers at the monthly meetings. The problems solved at weekly meetings are not the type that requires decisions at the corporate level. The problems that are solved at weekly meetings must be solvable at the supervisor and middle manager level.

By definition, the meetings are held weekly, but there is no particular day that is preferable as long as each meeting is held on the same day. In most cases, the presentation is for investigations that were initiated the prior week. Some issues could drag on for a while, but they will eventually catch the attention of senior management and be given a higher level of attention. Still, some problems are more difficult to solve because they require support from senior management. The weekly meetings address issues that are within the capability of lower management to solve.

The Monthly Meeting

While it is important that senior management be involved in TM, and occasionally attend daily and weekly meetings, it is not necessary for them to attend on a regular basis. Senior managers have other duties and often travel a great deal. It is often very difficult to get time on their calendars, but getting the senior managers together just once a month should be possible. In addition, it isn't absolutely necessary that the monthly meeting be conducted on the same day every month. Bottom line, if the managers really want to get together once a month, they will find a way. Usually, they turn the scheduling over to their administrative assistants.

In a monthly meeting, the middle managers present the top 5 Detractors and Enhancers to the execs and explain the progress that has been achieved. For those problems that will just not go away, or for the solutions that require executive intervention, the senior managers can provide the necessary support to reach a solution.

The monthly TAKT meeting is instrumental in getting the tough problems solved, but it is also a very effective way of celebrating. Middle managers that have a record for solving problems and improving their processes quickly get the attention of the executives. Continuous improvement becomes a way of life and employees at all levels are involved in the effort. When workers know that senior managers are involved in the continuous improvement effort, their motivation to get involved also is improved. They know that their efforts will be recognized and appreciated.

It's a Team Effort!

If everyone is not involved in implementing and sustaining TM, it will not be totally successful. I have seen partial and full involvement and believe me when I say that full involvement gets the best results! I have also seen minimal senior management involvement and TM still gets good results, but not as good as would be possible with full cooperation from the executives. The best results by far are achieved when all levels operate as a team – each doing what they can to create and sustain an atmosphere of continuous improvement.

With upper management involvement, the TM tool will continue to be effective. If management does not continue to be involved, the workers will notice it and the tool will eventually lose its effectiveness. The best way to continue to reap the benefits of TM is to achieve total involvement by all employees at all levels. Conversely, the certain death of TM begins with managerial indifference. It will continue to be vibrant and effective with full and exuberant managerial involvement.

Chapter 2 – Implementing TAKT Management

The most effective way of implementing TM is to implement it across the complete line simultaneously. However, with a large plant, the thought of a complete turn-on from front to back presents a very significant challenge. Simultaneous implementation of TM throughout the plant requires a great deal of planning and training, but it certainly can be done.

To say that one just "Turns On" TM is a bit misleading. TM can never just be turned on; it requires a great deal of planning and preparation. Of course, there is a great deal of training to complete – everybody must know what to do and how to do it. People must not only understand the mechanics of TAKT Management, they must understand what their responsibilities are and they must be advocates of the tool. Implementing TM not only requires a great deal of training, it requires a substantial amount of practice.

When TAKT Charts are introduced to the line, it takes some time and a great deal of training to make the workers aware of the reason for TM. They must understand the importance of absolute honesty in the completion of the charts. They must also understand that TM is not a report card. While it is true that the output is monitored, the important element of TM is the ability to determine what the Detractors to production are and to find solutions to those detractors. They must also understand that Enhancers are equally important because they tell us what was done to improve production. To achieve that necessary level of knowledge and understanding, the workers require considerable training and a reasonable amount of time to practice.

In one manufacturing plant, it took well over a month to complete the break-in period; in another it took only several weeks. The difference in the two situations was directly proportionate with the level of emphasis that was placed on the project by management – particularly senior management. In the first case, management was not demanding and did not press for the successful completion of the project. They did have an all employees meeting to show support for the project, but that is about as far as their exuberance went. Management received some training but did not completely

210

understand the dynamics of TM. As a result of all of this limited understanding, management sometimes made decisions that hurt the implementation effort.

In the second case, management showed their commitment to the effort by pushing the training and getting involved in it. They had a big kick-off ceremony and made a big deal of it. They walked around and promoted TM, telling the employees how important that it was to their future. Middle managers were asked to promote, not just accept the effort. Probably most important event in that scenario was that senior management received Fast Cycle Production training. It was not the same level of training that was given to the implementation team, but it was equal to the level received by the workers. The training that they received gave them a very good understanding of what their role was in the implementation effort. They understood what the entries on the charts should look like and were able to walk around and understand what was taking place.

Senior managers pushed their staff for their buy-in in to the effort and visited the floor fairly often before and during the implementation to show their support. The advanced PR worked and the employees got the message. Most important to me was the fact that senior management didn't just lose interest after the initial gains were realized. They continued to show their support long afterward.

As you can well imagine, there was a great deal of difference in the results of the two productivity improvement projects. The company whose management did not aggressively support the project did not achieve anywhere near the results of the company whose senior managers strongly supported the project and continued to show their support ... even long afterwards. Each company got results, but the company with the more supportive management got great results quickly and sustained the improvement effort over the long term.

When implementing TM, it is very important that employee involvement be given a high priority. It will not work to train a select group of people and have others that are either ignored, or given very little attention. The backlash will be significant when people are "left out". When filling in for the person responsible for completing the hourly charts, the untrained employees will put something on the charts that should not be there. It is so much better to train the complete department rather than just a select few.

People not only want to be involved in an improvement effort, they pretty much demand it. If you fail to involve each and every one of them, it will result in diminished results. The resistance will not be apparent, but it will be there. And it really doesn't have to happen if each employee is given quality training prior to the turn-on of TM.

I emphasize QUALITY TRAINING because too often I've seen the training conducted by people that still only had a superficial knowledge of TM. They understood the mechanics but were unable to give the employees a real understanding of TM. They teach the basic mechanics of how to fill out a chart, such as how often the entries should be made, or what the keywords should be for different types of Detractors and Enhancers. Again, they often fail to teach the **Whys** of TM, such as **why** accuracy is so important, and **why** the data is entered into the database. In addition, they quite often fail to teach the employees what happens to the data and how it is recovered, along with the importance of daily, weekly and monthly meetings.

In the defense of the Core Teams, (*A Core Team member is a person assigned to the project by the company who, after intense training in FCP tools and values, assumes responsibility for the implementation of Fast Cycle Production*) I must say that they did their best and some of them did a very good job of teaching. The problem goes back to the beginning of the project when the Core Team people were selected. At that time we were looking for potential senseis that would teach and mentor. When asked by management what kind of people that I would like to have on the implementation team, I would respond with the following requirements:

1. They should be highly respected by their peers and management for their job knowledge.
2. They should be well spoken and able to communicate well.
3. They should have good presentation skills.
4. They should be willing to teach.

Now the old saying that 3 out of 4 isn't bad doesn't apply here if the person does not have teaching skills. In fact, I believe that the applicant should possess all 4 attributes. I think that the company management has the notion that each core team member will just go back into his or her old jobs once the project is finished. They look around for someone who is "free" right now and who has some of the necessary attributes.

Even though I was never given a "perfect" team, I was fortunate in each and every case to get teams that were exuberant and that worked very hard to make the project a success. There was one project when I was given a large team to work with that had 8 members, 3 of which were really not suited for core team duties. They worked hard, were wonderful people and I enjoyed working with them. I just wish that they had all been better teachers with good presentation skills. A few of them really didn't want to be on the team because they were more comfortable in the role of the individual contributor. The team would have been stronger had all of the members been fully qualified for the team.

Implementing TM in a plant is no easy task and the team assigned to put it all together has to do a great deal of training if the project is to be a success. In a large plant it means that the training will sometime involve hundreds of people. Remember, it is not enough that just production people be trained, support people have to be trained also. This means that the core team and the consultant should all participate in the training. In most core teams that I have been given, only one or two people conducted the training because the others didn't feel comfortable teaching the material. That meant that the training for some of the production and support people was "delayed" or was not completed prior to implementation.

When TM is planned and implemented properly, it will provide a company with the best continuous improvement tool possible. It will help that company to reach productivity levels once only dreamed of. If sustained, it could help that company reach world-class levels. It can help production find and open its constrictions. If you can open your bottlenecks, you can improve your output capability. If you do not open your constrictions, you will live with whatever output capability that you have until you do.

Tom Clason

Part 4 – Modern Constriction Theory – Gate Busting!

Tom Clason

Chapter 1 – Basic FCP Constriction Theory

There has been a great deal written about constriction theory and how to deal with constrictions in production. The most notable author in my mind is Eli Goldratt, who I believe really made the study of constrictions popular. As I said earlier, Eli was the man who shaped my way of thinking when it came to constrictions. He gave me a new perspective on how to deal with them.

There is one thing that is true about the constriction, and that is that it determines the output capability of a factory, a department or a production line. If every single department except one can produce 100 widgets every hour, the factory cannot produce 100 widgets every hour. They can only produce what that one bottleneck department can produce. If that department is only able to produce 90 widgets an hour, that will be what the factory produces.

You can spend a lot of money in the other departments and you will not affect the total output of the factory. You can also spend a lot of money needlessly in the bottleneck department. But if you want to effectively deal with bottlenecks, you will look for the Right Hand Simple solutions to your bottleneck problems.

You must open the bottleneck in the slowest department and raise its output to the required level if you are to raise the factory output. To raise the output of the bottleneck department, you must find and open the bottleneck within that department. Bottom line, every plant, department, or work center has a bottleneck, even the bottleneck department. Open the bottleneck and you increase output. When that bottleneck is opened, you go to the next bottleneck and work to open it and keep going ad infinitum. It really is as simple as that. I call it "Gate Busting". But first, we must find out **why** the bottleneck is the bottleneck.

Is The Bottleneck Really The Bottleneck?

Usually, when we think of a bottleneck, we think that the problem with it is a machine that can't put out enough widgets, or a work center that doesn't have enough people to get the job done. We might also think that inexperienced workers or even unmotivated workers cause the problem or that the supervisor is incompetent. While any one of the reasons given could be true, the real reason for the bottleneck probably doesn't have anything to do with people or machines. Before we can fix the bottleneck, we have to find out exactly what is causing the problem.

Other than the reasons already mentioned, what other factors can create a bottleneck? Well, there are quite a few of them. Although they are recognized as "problems", they aren't often given credit for creating the bottleneck. They are usually considered little irritations that we have to work through. Here are some examples of reasons that are often overlooked:

Quality or Yield

When a manufacturing work center feeds a test or inspection department, bad quality could create a huge problem. Every time a widget fails inspection or test, there is additional work to accomplish, which could include paperwork and rework. Whenever additional work is required, product could very well back up. It appears to everyone that the test or inspection department is having problems, but the quality problems are really created upstream from them.

In some plants, if quality problems are detected, the complete batch may be returned to the previous work center where the problem will be analyzed and a determination made for the handling of the defective product. When defective product is returned to a department it becomes a resource drainer. It inhibits the ability of the department to meet their output requirements because people have to work on the defective product. They either have to perform rework duties or scrap the defective product. In either case, it takes time away from normal duties.

But the problem could also be associated with wasteful process steps in the test or inspection departments themselves. The bottleneck may be the test or inspection department. In fact, productivity is not always a priority of test and inspection departments, although it should be. There could be excessive paperwork, software issues, training problems or even transport problems that inhibit productivity and create the bottleneck.

There is also often a problem of product being held up for long periods of time because the quality engineers have not determined either the reason for, or the source of certain quality problems. When product is held up for such issues, it will probably mean that an order misses its due date. Quality issues are costly because they present the kinds of problems that are difficult to solve. Product quality can be, and often is the reason that a department is the bottleneck even though that department is not responsible for the bad quality.

Machine Setups

It is a fact that setups are a problem in manufacturing, but it is also a fact that they are pretty much accepted as necessary evils. Long setups are often countered with large batch sizes or they are merely accepted as a fact of life. If it were suggested that the setup time could be reduced, the response may well be that reducing setup time would result in reduced quality – a ridiculous statement.

Setups take time away from production. The longer the setup, the more output lost. Reduce the setup time and output improves. The more that the setup is reduced, the more output that you will recover. It is a very simple formula, so then why is setup time often ignored? In my opinion, many companies just accept the setup times as a given and increase their resources rather than aggressively attack setup times through setup reduction. The truth is that **all** setup times can be reduced by at least 50%. In many cases, 50% is just a drop in the bucket.

Getting back to the point of understanding bottlenecks, let us assume that setup time is the reason for a bottleneck simply because too much time is being lost doing setups. Put another way, too much production time is lost because it is spent doing setups. If you accept that assumption, and the assumption that something can be done to reduce setups, then you must understand that a bottleneck can be eliminated through setup reduction.

Maintenance Issues

When machines break down, they create production problems. When they break down often and unexpectedly, they can create a bottleneck. Little nit-picking problems that just keep popping their ugly heads up and grabbing you when you need it the least can quickly cause the widgets to back up. Pervasive maintenance issues on equipment can be much more than an irritation, although they are often treated as such. A major machine breakdown will get a lot of attention, but the little problems just manage to eat your lunch day after day and they are hard to explain because operators just try to work through the problems and eat the loss.

What is surprising to many people is just how much output is lost because of little, mostly unreported machine problems. They can very easily create a bottleneck and never be suspected as a problem. Even more problematic is the fact that they are often very difficult to fix because the root cause of the problem may be associated with a number of factors: machine age, operator training and, of course, poor preventative maintenance. A faulty maintenance policy could also be the cause. It could be that maintenance is so understaffed and busy that unless the problem is a big one, you can forget about support, or maintenance response time is so long that it contributes to the output loss. Whatever the reason or reasons, it is true that maintenance problems can create bottlenecks but aren't always identified as the reason for a bottleneck. To ignore them as a possibility is not a wise policy to follow.

Whenever a machine is a standalone machine, meaning that there are no other machines like it, and most or the entire product on the line passes through that machine, a maintenance issue on that standalone machine could create serious output problems. A standalone machine with maintenance problems can very easily become a bottleneck, even though it may have a substantial catch up capability under normal circumstances. Unless there is an alternative route for the product to follow, the product will build up at the standalone machine when it is struggling with maintenance issues. Having a standalone machine at a critical production point is not a wise policy unless there is no other alternative.

Long Term Overtime

OK, I'll bet that you're eyebrows are rising a bit. How could sustained overtime create a bottleneck you ask? Well, to tell you the truth, nobody really knows why, but students of the science of manufacturing know it to be true. When overtime is sustained for long periods of time, a loss of productivity occurs. Now we're not talking about a little overtime here and there, we are talking about serious overtime that includes the weekends.

When a work center or department is having trouble keeping up with the demand for product, overtime is usually the solution that is used to solve what is usually considered at first to be a short-term problem. When overtime is held to reasonable levels (eight to ten percent) overtime is a good choice to fill the deficit. Of course, while overtime should be used as a temporary solution, it is often used as the **only** solution. The correct course of action should be to find out what the real reason for the output deficit is through process analysis or TAKT Management, and work to open the bottleneck. Working to improve process capability is the right road to follow. Without preparation, a company can only throw overtime at their crisis output problems.

When a department has to work serious overtime day after day and week after week, its productivity will suffer. All of a sudden the department is producing in seven days what they used to produce in five. Sounds crazy? Well, it happens, and the reason for it could be one or all of the following:

1. The employees are fed up with overtime and are pacing themselves.
2. Employees may not like the overtime, but they have adjusted to the amount of money they are now making and want it to continue.
3. Workers have lost hope that the overtime will ever end and have no interest in trying to be productive.
4. Workers and management are worn out from working so much that the energy level is low.
5. Workers and/or management may be getting pressure from wives and husbands to work less and become disgruntled because the overtime is mandatory. (Common)

You can pick any of the above or you can come up with reasons of your own. If you have worked endless overtime, you probably do have some theories of your own. If you are completely honest, you know that the following scenario, or one very similar takes place:

- When overtime is first initiated, everything runs well. Output is good and the orders get out on time. People like the extra money.

221

- After a month or so of lost weekends, people start calling in sick in order to get time off. Absenteeism increases.
- Output drops off and production problems seem to increase, resulting in missed orders.
- Quality problems increase.
- Morale is suddenly a problem.
- Management puts pressure on production to increase output. Stress levels build up with the workers and management.
- Overtime becomes a way of life – Workers and management see no end in sight.

Think about it – if you were in a situation where you had no personal life you would long for one. If a weekend with your family were something to dream about you would get pressure from your family. If your workday lasted from breakfast to bedtime, you would be looking for a way out. Are these scenarios common? You bet they are when a company has a big demand for their product and they don't have the ability to meet the demand. If an order is taken and passed on to manufacturing, then manufacturing will try to get that order shipped any way that it can. Process mistakes are made and unfortunately, excessive overtime is one of those mistakes. But it is not just a mistake in judgment; it is a mistake in judgment that carries a heavy price tag. Overtime is expensive and it can eat up profits big time!

Companies that do not prepare for the possibility of increased orders without increased capabilities will most likely turn to overtime to solve what they consider a temporary inability on their part to meet lead times and order quantities. Suddenly they find that, after a few months when the situation has worsened, what was once temporary is approaching permanent.

Overtime works well when it is limited and occasional. The best way to avoid falling into the overtime trap is to prepare the process for increased capability. Improve process capability in advance of the crisis and the overtime may certainly hit when the big orders start coming in, but the recovery will be quick and the overtime will go away quickly.

Give Me What I Want When I Want It – With The Quantity I Need!

One of the more insidious creators of bottlenecks is the inconsistent delivery of product from work center to work center. When one department feeds another, the customer is pretty much at the mercy of the supplier when it comes to the delivery of product. When TAKT Management is in place, this condition will not exist. When TM is not in place, there could be problems with product delivery. I have seen product delivery that is so inconsistent that almost as much time is lost because of shutdowns due to the fact that there is nothing to work on, as there is production time.

It is always costly to run out of work to do, especially if the work center has trouble keeping up with a steady delivery of work. If that work center is a bottleneck, it is impossible to catch up unless overtime is used. Even with overtime, the bottleneck area will have a tough time making up the lost time. This brings up the truths about constrictions:

1. **OUTPUT LOST AT THE BOTTLENECK CANNOT BE MADE UP AT NORMAL COST!**
2. **NEVER BUILD UP EXCESS WIP IN FRONT OF THE BOTTLENECK!**
3. **THE LINE CANNOT PRODUCE MORE THAN THE BOTTLENECK!**

Face the fact that the slowest work center in the factory or department does not have a catch up capability. How could they? They are already the slowest of them all. How can they possibly catch up? Obviously, they cannot. So when it is said that output lost is lost forever at normal cost, it is an indisputable fact. If the bottleneck runs out of work, it cannot continue to put out product. The best way to keep the bottleneck from running out of work is to deliver the amount of work needed when it is needed. Failing to feed the bottleneck regularly results in output lost that is lost forever. It cannot be made up at normal cost. If the company is working 24/7, the output is lost forever - period!

There are other bottleneck creators, such as excessive paperwork and other non-value add tasks that are added to the normal duties of a work center that inhibit production to such a point that they actually turn a productive department into a bottleneck. The important thing to remember is that you must first determine the **true** reason for the bottleneck.

Chapter 2 – Gate Busting – How to Deal With The Constriction

The best way to deal with the constriction is to **HELP** it. Anything else that you do will be insignificant in comparison. Department meetings and managerial pressure sessions will not do a thing for you. If you want to get more output out of the bottleneck, HELP it. There is not a single bottleneck out there that can significantly improve its output without help. Pressure never produces anything other than temporary improvement.

When I say, "Help the constriction", I am not necessarily saying that you should pour money and/or resources into it; I am saying that you should go in, find the root cause and fix it. Help the bottleneck to reduce its cycle time and increase its output. Help the constriction by being **F*A*S*T*E*R!**

F*A*S*T*E*R – Help For The Constriction

- **F**ind the bottleneck
- **A**nalyze the process
- **S**implify – Remove the waste
- **T**rain – Workers and management
- **E**valuate – Test the results
- **R**ecognize – Recognize the workers and Celebrate

With **F*A*S*T*E*R**, you have a system that will provide help to the bottleneck(s). Now, instead of putting more pressure on the bottleneck, you can give it the help that it really needs. What should be noted is that the bottleneck department is often asked to find its own solutions. It is kind of crazy to ask the bottleneck to provide resources to work out its problems because the bottleneck usually needs every resource at its disposal to just survive. For that reason, the best way to help that bottleneck is to provide the resources and make it **F*A*S*T*E*R.**

Find The Bottleneck

Remember to make sure that you know the reason for the bottleneck. When you know the reason, quite often the solution is intuitive. Also remember that whenever a bottleneck is eliminated, another will spring up to take its place. Find out where the new bottleneck is. In many cases the new constriction becomes apparent. The slowest point in the process will often stand out. But finding the new bottleneck may be a little difficult because the outputs of the individual work centers may be very close. There may not be a glaring difference between their output capabilities. When there is not a big difference, TAKT Management may be the best way of finding the new bottleneck.

TAKT Management is a great way of finding the bottleneck department and may even point to the process step that is the true bottleneck. Through TAKT Analysis, the constriction point(s) can be identified. Daily, weekly and monthly meetings will help to find the constriction points also. When detractors are identified and eliminated, the result could be the opening of a bottleneck. TM is a great Gate Buster!

Since there is always a WIP buildup at a constriction point, line of sight management can be extremely helpful in pinpointing the location of the constriction points. When you find the buildup, look at the TAKT Charts to see why the department is having trouble. Look at the average output of the bottleneck. If everything is running normally and there are no outstanding problems, and still the department is not keeping up with the output of its supplier, it is a strong indication that the work center with the WIP buildup is likely to be a bottleneck.

Analyze The Bottleneck

Once the bottleneck is located, we next have to find out why the bottleneck is the bottleneck. The answers may not be obvious, we will have to do some analysis to find out the reasons. There are three great tools that will tell us all we need to know about the bottleneck:

1. TAKT Analysis
2. Process Mapping (flowcharting)
3. Simulation Modeling

TAKT Analysis can be instrumental in uncovering the hidden reasons for a bottleneck. Constrictions can have a variety of causes such as excess paperwork, resource issues such as absenteeism, product quality or even the attributes of the product itself. Certain types of product can present huge difficulties to a work center. That same product type may not present problems to all production work centers but could kill one single work center because it presents some complexities that affect only that one work center. On the other hand, some product types are demons to produce in virtually all departments. In addition to being difficult to produce, some product types present manufacturing with low yields, high scrap rates and rework problems.

TAKT Analysis will show these kinds of product related problems because they will appear as Detractors on the TAKT Charts. Through TAKT Analysis you can also determine the quantitative effect of the problem. You can look back to determine when the problem started and how often it has hit. TAKT Analysis can be a powerful ally in constriction analysis.

Process Mapping is a powerful constriction analysis tool. Many constrictions are the result of an accumulation of wasteful process steps that inhibit a department's ability to meet output requirements. Those little wasteful process steps are like a cancer that eats into a work centers production capability and just go undetected. Nobody knows what is killing their productivity because they take these little duties for granted. Process mapping exposes them because if done correctly, it documents every single little task that has to be completed by the workers. When some of the little tasks that inhibit output just do not make sense and are wasteful, process mapping will expose them.

Simulation Modeling is the perfect complement to Process Mapping for constriction analysis. With simulation modeling you can do multiple what-ifs that show what the process could do if certain changes were made. It can also show the effect that the changes would have on downstream operations. It will not only show how much more product could be handled by the department, but it will show how much the company could produce if the bottleneck were removed. The real value of simulation modeling cannot be overstated, but it must be used properly.

Simplify The Process

Once the process has been analyzed and the alternatives identified and tested, the next step is to implement the changes that will bust the gate. The first step in implementing the changes is to make sure that **everyone** that will be affected by the changes is involved in the implementation of the changes, not just told what to do. They have to know the reasons for the changes, what they will accomplish, and what effect the changes will have on other areas both upstream and down. In addition, everyone must know what exactly what their responsibilities will be in the implementation process. Finally, the procedural changes must be documented, preferably through process mapping and process run rules written to cover all what-if scenarios.

Train The Appropriate People

Whenever changes are implemented, the people that are directly responsible for working the changes on a day-to-day basis must be trained. If the changes involved removing wasteful process steps, the workers should be involved in evaluating the solutions. They are not given veto power, but there is good reason for their involvement - Their involvement helps to secure their buy-in for the solutions. By securing the buy-in of the workers you have just increased the probability of the immediate and long-term success of the solutions. Make the workers part of the solution!

Evaluate The Results of The Changes

Once the changes are implemented and the constriction is opened up, your work has not ended yet. What has be done can be undone if there is no follow-up work. You must make sure that the improvements have been sustained and the new procedures followed. It is the only way to make sure that the new process is working as planned.

If there is no improvement after the changes are implemented, you must find out why. First off, does everyone really understand the new procedure? Second, Is **everyone** on **all shifts** trained? If the training was insufficient or incomplete, you may need to go back and retrain the appropriate people. Third, are the workers really convinced that the changes are good and are they advocates of them? If they don't really believe the changes are for the better, the workers may not really support them. Getting the workers support is important. If the workers are convinced the changes are worthless or counter-productive, or just not convinced that they are worthwhile, there may be negative Pygmalion responses to the changes. The workers may sub-consciously work to make the changes unsuccessful. A little bit of work to get their buy-in will help to make the implementation successful.

<u>R</u>ecognize The Accomplishment

When the constriction has been opened and cycle time and output have improved, it is time to celebrate! You shouldn't wait until you are convinced that the changes will stick, you should celebrate quickly and enthusiastically. In your celebration, you make sure that everyone (workers and management) knows what has been accomplished, while giving the credit for the success to the workers who made it happen. The worst thing that can be done is for management to celebrate the victory first. The first celebration should involve the workers. There will be plenty of time for management to celebrate later.

Celebrating need not be expensive. In fact, the best methods of celebrating are very inexpensive. A great way of celebrating is to put up what I call "High Five" charts around the plant showing what was accomplished and the names of the workers involved in the success. Make sure that the off-shift people are represented also. In one plant, the charts showed the pictures of the workers that made it happen. These kinds of celebrations are highly effective and help to make successful changes last indefinitely. Recognizing the accomplishments is as important as any element of **F*A*S*T*E*R** yet is the one most likely to be excluded.

Management should publicly recognize the workers that took part in the implementation of the changes and who made the improvements successful. Workers enjoy being recognized by management and peers. Celebrations should never be disregarded or delayed regardless of the magnitude of the success. When gates are busted, everybody benefits!

Part 5 - FCP Setup Reduction – The Team Method

Tom Clason

Chapter 1 – The JIT Way

Shingeo Shingo proved that setup reduction was not only possible, but was the right thing to do. He proved that setup reduction could be accomplished on virtually any machine and that setup times could be reduced from hours to minutes, or even a single minute.

Dr. Shingo's book, "A Revolution in Manufacturing: The SMED System" was one of the more outstanding innovative elements of JIT. As one of the developers of JIT, Dr. Shingo's approach to a long-standing problem in production and the approach that he took to solving the problem of excessive tool changeover times was the work of a genius. Although he acknowledged that it was important to cut setup times, he maintained that it was also important to "diminish lot size, and even loads simultaneously".

In the minds of the JIT architects, of which Dr. Shingo was a major contributor, establishing and maintaining a quick and lean manufacturing process was essential. It was not enough to improve output; you must simultaneously keep WIP levels low and stable throughout the process. Dr. Shingo maintained that increasing batch sizes to improve machine utilization was counter-productive. He was clear that reducing setup times while reducing batch sizes provided the producer with the proper solution; in that machine or factory utilization and efficiencies are maintained (of critical importance for gating operations) without compromising factory WIP levels.

But reducing batch sizes must be accomplished intelligently. If the batch size is reduced too much without a reduction of the setup time, it could result in a significant loss of output. The appropriate reduction of setup time accompanied by the proper batch size reduction will result in maximized output and equipment utilization. Figure 5.1.1 represents a data matrix showing the relationship between batch sizes, setup times and output.

The chart in Figure 5.1.1 assumes a 24-hour day (1440 minutes), a machine output rate of 1 unit per minute or a maximum of 1440 units in a 24-hour day. Each batch takes as many minutes to process as the amount in the batch. For example, a batch size of 100 will take 100 minutes to complete. After each batch is completed, a new setup is completed. The setup times range from 10 to 60 minutes in increments of 10 minutes.

Notice that with a batch size of 100 and a setup time of 60 minutes, the maximum output for a 24-hour day is 900 units. This means that there is an output loss of 540 units ($1440 - 900 = 540$) due to setup time. That represents a loss of 37.5% - not an insignificant amount. If the batch size was reduced to 50 units and the setup time remained at 60 minutes, the output would fall to 655. That would represent an output loss of 54.5%. In this scenario, more time would be spent doing setups than producing product. Obviously, this situation wouldn't be a good one to be in.

But the batch size problem is still not solved. Obviously, there is a loss of output when batch sizes are reduced. Figure 5.1.2 shows a chart that gives you a quick comparison of the output loss for each of the setup times and associated batch sizes. If an output of 1100 was needed (A 22% increase) to meet output requirements and the present output was 900 with a batch size of 100, and we reduced the setup time by 50%, we would be able to meet our output requirements but we would not be able to reduce our batch size. If our goal were to reduce our batch size by 30% **and** increase our output to 1100, we would have to reduce our setup time to 20 minutes (66.7%).

Knee of The Curve

Batch Size				Setup Time		
	60	**50**	**40**	**30**	**20**	**10**
100	900	960	1029	1108	1200	1309
95	883	943	1013	1094	1190	1303
90	864	926	997	1080	1178	1296
85	844	907	979	1064	1166	1288
80	823	886	960	1047	1152	1280
75	800	864	939	1029	1137	1271
70	775	840	916	1008	1120	1260
65	749	814	891	985	1101	1248
60	720	785	864	960	1080	1234
55	689	754	834	932	1056	1218
50	655	720	800	900	1029	1200
45	617	682	762	864	997	1178
40	576	640	720	823	960	1152
35	531	593	672	775	916	1120
30	480	540	617	720	864	1080
25	424	480	554	655	800	1029
20	360	411	480	576	720	960
15	288	332	393	480	617	864
10	206	240	288	360	480	720
5	111	131	160	206	288	480

Figure 5.1.1

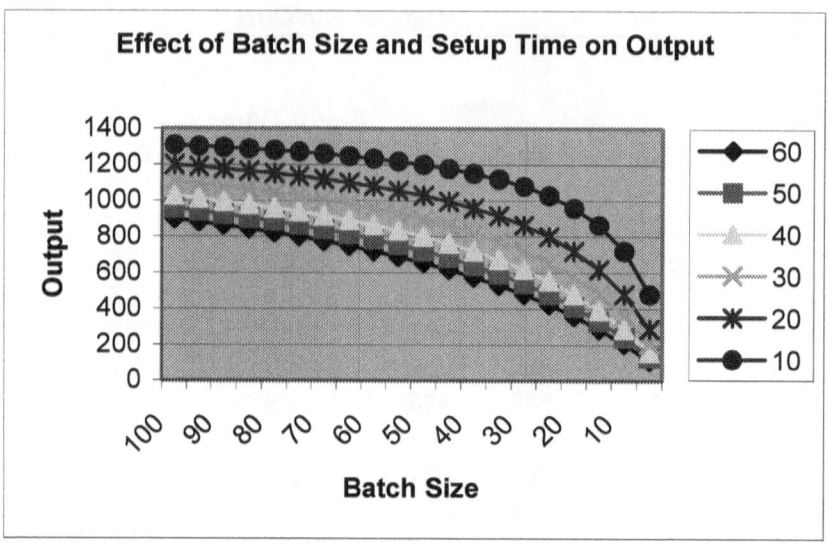

Figure 5.1.2

Notice that with a setup time of 10 minutes, the output loss is insignificant moving from a batch size of 100 to about a batch size of 50, after which the output falls almost exponentially. With a setup time of 60 minutes, the output drops off significantly almost immediately as we reduce batch sizes. With a setup time of 60 minutes, if we moved from a batch size of 100 to a batch size of 50, the output loss would be 245 units. With a setup time of 10 minutes, the same reduction in batch size would only result in an output loss of 109 units. This lesser impact to output capability may be quite acceptable for your factory based on your customer demands and is clearly more favorable than an output loss of 245 units. At the same time, cycle time is improved as the smaller batch size helps to reduce the overall WIP in the factory.

The 3 Stages of SMED (Single Minute Exchange of Dies)

The SMED setup reduction system is powerful, and consistently effective when applied properly. It is brilliantly simple and easy to follow. The successes that Dr. Shingo experienced with setup reductions can be realized in any industry. There are 3 major stages of the SMED setup reduction process. Each stage is designed to either change the order of the setup or to remove waste from the setup procedure. The major emphasis is placed on converting internal steps to external steps. First, a simple definition of what constitutes an internal or external step:

External Step: A setup step that can be completed while the machine is still running.

Internal Step: A setup step that must be completed while the machine is off.

Stage 1 – Separating internal and external setup as it exists. Every step of the setup is identified and studied to determine if it is either an internal or external setup step.

Stage 2 – Converting internal setup steps to external setup steps. Here, the order of the setup steps is altered. Instead of completing the setup step after the machine has been disabled or turned off, the step is completed prior to the machine finishing the last part of the previous batch.

Stage 3 – Streamline all aspects of the setup operation. The order of the internal and/or the external setup steps could be changed to improve the efficiency of the setup. The best setup streamlining would be to remove setup steps that add no value to the setup, reducing the amount of distance that the person(s) doing the setup have to travel (circles of motion) or even modifying the machine itself to make it easier to perform certain setup steps, such as changing free turning calibration dials to click type calibration dials. Also, modify the machine to make it impossible to make a mistake when inserting fixtures or plates. The inserted pieces can only be successfully placed one way and one way only.

Dr. Shingo did not just make one attempt at reducing setup and then giving up, he kept attacking until he was convinced that the setup could not possibly be reduced further. He was tireless in his attack on reducing setup time. In 1969 he worked with Toyota Motor Company to reduce the setup time on a 1000-ton press. He was successful in reducing the setup time from 4 hours to 90 minutes in six months time.

The fact that it took 6 months to get the setup time down to 90 minutes is proof that they did not just try once and give up. Even though they were happy about the initial success, management insisted that it was necessary for the setup time to be reduced to 3 minutes or less. Using Dr. Shingo's method of converting internal steps to external steps, they were able to reduce the 90-minute setup to 3 minutes. Getting the setup time down to three minutes took another three months. This was the birth of SMED.

Using the SMED procedures, Toyota and other Japanese companies were able to reduce setup times from hours to minutes. These kinds of results were not rare occurrences. Dr. Shingo had started a mini-revolution with his discovery of SMED. It, like other right hand simple solutions, was brilliant in its ease of application and the results that it achieved. Even today, SMED is followed religiously when setup reduction is needed.

Chapter 2 – The FCP Approach To Setup Reduction

Setup reduction is a major element of FCP. Setup reduction can open constrictions, reduce cycle time and improve output. The savings from setup reduction are usually huge. Like other elements of FCP, setup reduction is not only appropriate for manufacturing; it is also applicable in other areas. I used Setup Reduction procedures to improve the performance of the printing department of a major bank in California. I also used Setup Reduction principles to improve the process for implementing Engineering Change Orders in the engineering department of a medical instruments manufacturer. I even used Setup Reduction procedures to improve and reduce the time that it took to change shifts with several companies.

How do you improve on SMED? Well, for one thing, you certainly don't change the basic tenets of SMED. The foundational elements of SMED were solid when first used at Toyota and they are just as useful today. The results achieved were almost mind-boggling in their magnitude. Reducing setup time by 50%, which would be meaningful to most companies, was just a drop in the bucket to Dr. Shingo. Obviously, six hours to three minutes goes well beyond 50%, but was commonplace for Dr. Shingo. Huge reductions are only possible when you truly believe they are not only possible, but also quite probable and a true business imperative!

FCP demonstrates that traditional SMED procedures can certainly be applied a little differently for the modern day manufacturer. Having a brilliant engineer like Dr. Shingo leading a setup reduction effort would certainly be advantageous, but many companies are either short on JIT savvy engineers or have no industrial engineers on staff.

Some of the companies that I worked with had industrial engineering departments, but all of the engineers were tied up on machine sizing or standards. In one company, very few of the people in the IE department had IE degrees. In another company, a newly hired industrial engineer would spend a few months learning the process and then would become a supervisor on one of the lines.

Subsequently, the IE's were hired for supervisory positions and therefore were not available for setup reduction projects.

In one company that I worked with, a few IE's were responsible for setup times and had established setup procedures for all of the machines in the plant. The IE's did not communicate with, or involve any person other than the operator(s) that worked the machines, and only to instruct them in the setup procedures that they should follow. The process was totally autocratic. The operators were seldom, if ever asked for their input. In yet another company, the setup procedures were actually provided by the company that made the machine.

In each of these companies, the setups were excessively long and most disturbing of all, were accepted as facts of life. "The setup for this machine is eight hours long," I was told with conviction, "and we have done everything possible to get it down to that amount of time." Management had accepted that as fact and never challenged it. The productivity loss was staggering and, amazing as it sounds, management was looking for "other" ways to improve their output.

In each of these companies I promised that their setup time could be reduced by at least 50% - guaranteed! Heck, from my perspective, and considering what Dr. Shingo was able to do, and the results he was able to achieve, 50% was a walk in the park. In each case I was looking at setup times that were at least 5 times more than they should be.

In each case I was presented with a common obstacle: In order to reduce setup time, I had to rely on cooperation from the same people that created the prevailing setups. How could I get them to agree that they had established procedures that were at least twice what they should be? To tell the truth, when dealing only with the creators of the setup procedures, I had a great deal of difficulty getting them to work beyond their paradigm. They seemed convinced that the original creation was perfect. I had to find another way.

In my desperation, I came up with the original version of the team method of setup reduction. I convinced my client to allow me to put together a team of individuals that would work as a team to accomplish what was once a function of a single person. Like many solutions that are born of necessity, this was no exception. From the first day that the team method of setup was born, I have attempted to improve upon it.

Fortunately for me, I received a great deal of help from the people that I worked with. The teams that I trained were innovative and came up with ideas that worked so well, I took those ideas to the next job. From job to job, the system became more effective.

The improvements that were made were associated with how the teams were formed, what kinds of people should be included in the teams and how the setup should be recorded. Other improvements, which will be explained shortly, are designed to ensure that the new setup procedures stay in place, once implemented. I found that it is one thing to come up with a new and improved setup procedure, it is quite another to see it working, or even improving six months or a year later.

Phase 1 - Forming The Team

First of all, you cannot expect a team to just happen. Somebody has to form these teams, train them and facilitate the meetings. That same knowledge source will even be responsible for making sure that all of the follow-up activity occurs. The person who is given this responsibility should be thoroughly trained and knowledgeable in the SMED procedures as well as FCP principles and values – a sensei, if you will.

He or she should have many of the attributes required of a core team member: Respected by management and peers, good communicating skills, willing and able to train others and good leadership skills. The first time that a semi-trained or improperly motivated person is put into a position of setup team facilitator, you will see little or no meaningful results.

The selection of the team should be accomplished very carefully and should consist of most or all of the following people:

- Highly skilled operators from each shift. If there is only one shift, there should be at least three operators from that shift. These will be the people that teach their fellow operators how to complete the new setup. One of these operators will be the person who is filmed in the "before" and "after" scenarios.
- An operator from the preceding and/or following work centers.

- One or two Engineers. The engineers selected should be working engineers that preferably support the machines in the department.
- One Maintenance person. Again, the person selected should be the same person that supports the machines in the department.
- Quality engineer or technician. Having a person from the quality department gives credence to the new setup procedures.
- Area supervisor or manager – hopefully not a power-oriented person that will try to dominate the session or dictate, but one that will show support for the effort and work during the session as a team member.

The facilitator should get firm commitments from each member of the team to attend each and every session from beginning to end. It is highly disruptive to the team if team members arrive late or leave early. Each person assigned to the team should clear his or her calendars for the scheduled meetings. The facilitator should work carefully with each team member's calendar to ensure that the scheduled meetings do not create calendar conflicts.

Every single member of the team is critical to its success. If a person is likely to be disruptive or is openly not in favor of working with the team, they should not be included. The absence of this type of person will only benefit the team. Management should make it very clear that it supports the setup reduction effort and encourage cooperation from all of its reports. The facilitator should also copy management on all relevant email messages and replies to team members.

Phase 2 – Filming The Setup

First of all, you have to have a good camera that will show the elapsed time of the setup on the screen during playback. Some cameras will show the elapsed time while filming, but will not show the time during playback. This is important because the start and stop time of each setup step is recorded during the evaluation session. It is also recommended that the setups be recorded with a color camera.

The camera should be the type that can be hooked up to a large monitor for easy viewing by the team during evaluation. The monitor should be at least a 21-inch monitor so that the team members can easily view the details of the setup.

The next step is to choose the operator that will be filmed doing the setup. Again, the operator should be one who is respected by peers and management and willing to be filmed. Some operators may be a little camera shy at first, so it is important that there be a dress rehearsal of sorts to get the operator used to the camera and comfortable being filmed. Make sure that the operator knows that the setup should be completed exactly as it is completed normally. There should be no performance before the camera that does not represent reality. There might be a few differences between the way one operator completes a setup and another, but the basic elements should be exactly the same as the operator does the setup normally.

The location of the camera during certain portions of the setup should be thought out prior to the main filming. Not only does the operator have to become accustomed to the camera, the person doing the filming (usually the facilitator) has to become accustomed to filming the setup. The facilitator has to be in the proper location to make sure that the camera has the right angle and that he or she does not get in the way of the operator when the operator has to walk from one location to another. The person filming should be able to anticipate the operator's movements.

After filming the complete setup, which includes the moment the last good part passes through the machine until the first good part of the new lot passes through the machine, make sure that everything that should have been captured on film was. If anything at all went wrong with the filming, it should be redone. The filming is essentially the "before" scenario that will be shown to management and others later. It would be wasteful to have the results questioned because the filming didn't go right.

Before showing the film to the complete team, the facilitator and the operator should review the complete film from start to finish to make a preliminary determination that all of the necessary steps were completed and to make sure that there were no issues associated with the filming. Once the validation of the setup is complete you are ready for the next phase.

Phase 3 – Team Evaluation of The Setup

Prior to the setup, the facilitator should have the charts ready that will be used during the evaluation phase. The charts will record every single step of the setup and will be the basis for the improvement phase. The following is an example of a Setup Reduction Chart:

Step Number	Start Time	Finish Time	Duration of Step	Description of Step	EXT.	DEL.	Opportunity MOVE	MERGE /MOD.

Figure 4.2.1 – Setup Reduction Chart

The charts should be of sufficient size to be seen easily by the team members. A standard flip chart is the ideal size. The charts can be printed at a reasonable cost which facilitates the preparation of the charts. Here is an explanation of the chart columns.

Step Number
Each setup has a number of individual steps. Some setups have over a hundred different tasks that are required. Step 1 is the first step, Step 2 the second and so on.

Start Time
The time shown is the exact time that the setup step begins. The time shown is taken from the film time on the camera.

Finish Time
The exact time that the setup step is completed.

Duration of Step
The total time that it took to complete the setup step (Finish Time minus Start Time).

Description of Step
An accurate description of the process step.

Opportunity
The opportunity section is completed separately after the setup steps are entered and validated (Covered later).

Once all of the steps of the setup have been entered and validated, it is time to improve the setup. The team is ready to externalize as many steps as possible, which is the same as saying we want to complete as many steps as possible while the machine is still working on the preceding batch. Any steps that are wasteful will be eliminated. The setup may be rearranged to reduce time, and the team will look for opportunity to make machine modifications, which would improve the quality of the setup and even reduce the time that it takes to complete certain steps. But externalizing presents the greatest amount of opportunity in the early stages of setup reduction.

It is really necessary for the team to develop an aggressive attitude towards setup reduction. If they enter into the effort half-heartedly or in a lackadaisical manner, they will not be as successful as they would if they open their minds to the possibilities. You cannot expect results unless you really believe that you can be successful. Do not enter into a setup reduction effort with a mediocre goal in mind because you will be limiting your possibilities with Pygmalion expectations.

Completing The Chart

The chart is filled out in the order that the operator completes the setup. Each step is numbered and an explanation of each step, along with the start time, stop time and the duration of the step. This information is important because it makes analysis much easier.

After the chart is completed, the opportunity session will identify possibilities. Initially, the team only identifies possibilities; it is not yet the time to determine the feasibility of the possibilities. The possibilities shown on the chart are:

- **Externalize** – Can this particular step be completed before or after the machine is turned off?

- **Delete** – Can this step be removed without affecting the quality of the setup?
- **Move** – Should this step be completed earlier or later in the setup?
- **Merge/Modify** – Should the step be completed differently, merged with another step, or can the machine itself be modified to improve the quality and efficiency of the setup?

During the step-by-step opportunity analysis, the team does not attempt to arrive at a consensus. The question asked of the team is: IS IT POSSIBLE? If <u>one</u> person on the team believes that there is a possibility, the notation is made on the chart. The feasibility can be debated later, but initially only possibilities are explored.

Figure 4.2.2 shows a partially completed Setup Chart, complete with opportunity analysis. Only 3 steps of the setup are shown. A real setup could contain more than 50 steps. Notice that the setup starts at time zero, not the actual clock time of the setup. This makes it much easier for the team to determine what the duration of the setup is. If the clock on the camera only shows the actual date and time, the conversion is made when transferring information to the chart.

Step #	Start Time	Finish Time	Duration of Step	Description of Step	EXT.	DEL.	MOVE	MERGE /MOD.
1	0:00	4:30	4:30	Prepare new fixture	X			
2	4:30	4:45	0:15	Place new fixture beside machine	X			X
3	4:45	6:18	1:33	Remove screws on holding plate with screwdriver				X
4	6:18	6:45	0:27	Remove fixture holding plate and set holding plate on rack				X
5								

Figure 4.2.2 – Completed Setup Reduction Chart

Looking at the setup steps one by one, the following opportunities were identified by the team:

- **Step 1** – It was feasible that the preparation of the new fixture could be completed before the machine was turned off for the setup.

- **Step 2** – Since it was determined that it was feasible to externalize step 1, it makes sense to externalize step 2 also. The team also determined that there should be a holding rack for the new fixture located near or on the machine and that the new fixture be placed there in advance of the setup by someone other than the operator.
- **Step 3** – Using a power screwdriver can save time. Another possibility is to use screws that only require one turn to remove.
- **Step 4** – In this case, the operator had to walk about 30 feet to the rack. It was determined by the team that it was feasible to move the rack closer to the operator during the setup, then to move the rack to its holding location after the setup.

It is easy to see that more than 5 minutes can be removed from the setup in just the first 4 steps. This is not unusual. All setups can be reduced at least 50% and much more in many cases. Setup reduction is not rocket science; it is a right-hand simple solution. It doesn't cost a lot of money to do, it provides big benefits, and it is easy to implement.

Phase 4 - Create The New Setup

So far, we have filmed the setup and transcribed each step of the setup onto a chart for analysis. Each member of the team is empowered to identify opportunity and enter into discussion as to the feasibility of that opportunity. Team members discuss each step of the setup and come to consensus on the applicability of the changes.

Step by step opportunities are exploited and the setup is modified until a new setup is created from beginning to end. The new setup is then transcribed onto a new chart. Everyone involved in the effort should understand that the realized gains are only step one in the effort to reduce setup time. But the initial effort is not finished yet. The new setup must be practiced and fine-tuned.

Practice Makes Perfect

Each step of the new setup is now documented and it is time to determine whether the changes will really work. It is important that the new setup is not only quicker, but it must be a quality setup also. Quicker isn't always better when corners are cut that shouldn't be cut. The new setup must affect quality in a positive way.

Practicing the new setup will allow the team to fine tune the setup until it is perfect. Changes should be made and documented when appropriate. When the new setup is ready, it is filmed in the same way that the original setup was filmed. Filming the final setup along with the documented chart will be used for training other operators. The training of operators is key to the setup reduction effort.

There Is No New Setup Until <u>Everyone</u> Is Doing It

The team has created **the** new setup but so far only one or two operators know that the new setup is better than the old. Everyone even remotely associated with the new setup has to know that it is superior in every way and what the benefits of it will be. Celebrating the accomplishment of the new setup is very important and it will help to communicate the importance of the improvements to all employees, even employees in other departments. Since Setup Reduction is important to all departments, seeing the accomplishments of another department and how it is celebrated will make them more amenable to an identical effort in their own department.

The new setup cannot be optional. Management must make it clear that the new procedures should be followed to the letter. There could be resistance from some operators who feel that their own way of doing things is better, or that they just don't feel comfortable with change. The operators must come to understand the importance of their compliance. Once there is compliance, there can be sustained improvement. Just because the initial effort to reduce setup time is completed, does not mean that the final effort is completed. The setup reduction effort should be continued to include machine modification if necessary.

After The Setup Reduction – You're Still Not Finished!

The new setup has been created and it is a good one. Plenty of time has been cut and everyone is trained. The celebration has taken place and everyone, including management is happy with the results. The team has done its job very well and it is time to rest. You are partially correct, it is time to focus on another setup and maybe rest a little bit, but the real work with this setup has actually just begun.

All of the gains that you have worked so hard for can be lost quickly if you are not careful. Now it is time to establish an ongoing process to make absolutely certain that the operators continue to follow the setup procedure as it has been documented. The setup procedure must be audited regularly to ensure ongoing compliance.

The best way to audit setups is to do so on a no-notice basis and report your findings honestly. You do not want to stifle creativity by making it impossible for the operators to "find better ways" of doing the setup. If the operators think that they have found a better way, then everyone should be doing it that way. The operators should not change the setup procedure on their own, they should present their suggestions to the team and the team should evaluate the suggestions. The team should reassemble, film the new setup and compare the results with the approved setup procedure.

Bottom line, only the team that establishes the setup procedure should be able to make changes to it. If operators start to make changes, the gains that were realized will eventually be lost. Auditing, when done correctly, will ensure that operators follow the proper setup procedure on all shifts.

The auditing process should be taken quite seriously because the results of non-compliance could have a considerable effect on production. The reasons for setup reduction are to improve output and product quality. The results of the audits should be reported to all of management. Although management should not respond to a failed audit in a draconian fashion, they should require that the supervisor come up with an action plan to correct the situation. This could require additional training and/or counseling. It is not necessary to punish anyone, but it is necessary to correct the situation.

As you will notice later, audits are effective with other FCP tools. TAKT Management should use audits to make sure that the charts are timely and correct. Kanban and Dynaban Management can make good use of audits to keep Kanban and Dynaban discipline in place. Even established run rules should be audited on a no-notice basis. Shift changeovers should be audited regularly. In an FCP facility, efficient shift changes have a profound effect on output.

Any productive gain, including those realized by setup reduction can be lost through eroded compliance. Audits will prevent such losses and are very effective in a continuous improvement environment. The realized productivity gains are hard to obtain and they can be easily lost. Setup reduction can open constrictions. Audits can help to keep those constrictions open.

Part 6 – Kanban and Dynaban Management - The Right Hand Simple Solution That Still Works

Chapter 1 – Basic Kanban Theory

Work in Process, or WIP, buildup has always been a major problem in production. Every company knows (or should know) that it is expensive to have too much product on the floor, but many of them seem to have difficulty keeping their WIP under control. The most common method of dealing with late orders is to pump more product into production with the hope that it will miraculously be processed more quickly. This scenario has existed as long as I can remember and is still prevailing today.

Many companies feel that the best course of action is to rely upon schedules to control the WIP levels. After all, if the product is released to the floor in an orderly fashion, there should be no buildups. I refer to this type of thinking as traditional MRP or Push mentality. On the surface, the logic is fine, but in the reality of the production world, the logic is severely flawed.

Utopia does not exist in production because problems always crop up to spoil the day for the unprepared. The problems could be related to machines, process or people. Quality problems can crop up that can be difficult to solve. Training issues can create almost insurmountable problems that occur when productivity is most needed. Every time a company thinks that it has its problems under control, something will happen to burst the bubble. There has to be a way of controlling WIP while ensuring that adequate levels of WIP are present on the production floor.

FCP has tools to analyze the process and open bottlenecks. TAKT Management can identify problems that lead to solutions, but when problems crop up, there may not be an instant solution. The problem may take time to fix and meanwhile, new product continues to hit the floor as if everything was business as usual. When it hits the area experiencing the problem, WIP naturally starts to build up because the problem has slowed the output capability of the affected area(s). Failing to quickly recognize the problem at the release point and responding to it by cutting off the supply or fixing the problem, can cause the WIP to reach very serious levels. The temporary problem then becomes a much more serious one because the excess WIP that is piling up may become a bigger problem than the original one.

Kanbans provide the answer to effectively managing WIP on the production floor. The use of Kanbans originated in Japan during the JIT revolution and was designed to give each work center, from the release point to the last work center, the ability to signal to its supplier when it needed work and when to cut off the supply of work. The signal to start or stop work was based on a quantitative determination of what amount of WIP was needed to sustain production for any given work center at any given period of time. The solution was a brilliant right hand simple solution to a complex problem because it is easy to implement, is inexpensive and extremely effective.

In my studies of JIT and CFM, I have never found a WIP management tool that compares with Kanbans. They are uncomplicated and brilliant in their simplicity of implementation and their ease of use. Like all the other FCP tools, Kanbans are **always** effective when used properly. There are more expensive ways of controlling WIP, but you will not find a tool that is more effective.

The Difference Between a Queue And A Kanban

When a work center cannot process its product as fast as it arrives, there is a buildup of WIP. Conversely, when a work center is faster at producing product than its supplier, the work center could run out of work. Either case produces a situation that can adversely affect productivity and profit. Too much WIP can give the workers a sense of security, but it is false security, and quite unhealthy for the business. While too little WIP can be just as unhealthy as it can cause gaps in production and loss of output.

When work arrives at the input point of a work center, it is stored until it is needed. Without Kanban Management, it can just sit there if there are no controls in place. The product is essentially queued up. Without an intelligent Kanban discipline, there is no mechanism in place to signal the supplier when there is either too much or too little product.

A Kanban is much more than a WIP queue. First of all, Kanbans are strategically placed throughout a factory. Not every operation needs to have a Kanban; however, many operations can benefit from them. A Kanban has an upper and lower limit. The upper limit of the Kanban signals to the supplier that the supply of product to the customer should be stopped until the amount of product in the Kanban is below the established upper limit. The lower limit sends a signal to the supplier that the supply of product in the Kanban is dangerously low.

Normally, the flow of product from work center to work center is smooth and product coming in is processed as it arrives. TAKT Management will help to ensure the smooth flow of product, but TM in and of itself is not enough. Optimally, the release of product to the floor is sufficient to not only meet the needs of the line, but also to compensate for loss due to damage or quality defects. When everything is running smoothly, it is not difficult to predict the proper quantities to release to the floor. On the other hand, when problems exist, the difficulty factor increases exponentially.

It is very difficult to determine just how much product to release in order to meet requirements, especially in a complex production facility. Quality issues can crop up without notice, and all of a sudden orders are in danger of late delivery to the customer. Machine issues can block product from being shipped on time. Even people issues can slow or stop production. It is extremely difficult at best to determine the perfect release quantities all of the time. There are software tools out there that can help to determine the proper release quantities, but the best of them are extremely fallible.

To help compensate for the problem, the common solution to this dilemma is to release more than what is required to fill the orders. The extra amount is usually calculated based on historical values. Unfortunately, unforeseen instances can make the most careful calculation seem like a foolish mistake. As soon as the flow of product hits the problem area, WIP starts to build up. Without the proper controls in place, WIP will to continue to build up unless the problem is quickly resolved, and/or the work center that experienced the problem has an adequate catch up capability. If the problem persists, the WIP will continue to build up indefinitely if not brought into check. That is exactly what can and will happen without Kanbans … guaranteed!

Kanbans control the flow of product by shutting off the supply to the individual work center when the upper limit is reached. The supplier must then discontinue producing parts until there is a demand for more work from their customer. If the problem persists and the stoplight stays on for an extended period, the preceding Kanban could fill up. This domino effect will continue work center by work center until the initial release point is reached. Once that happens, no work will be released to the floor. An added benefit of this approach is that problems get addressed quickly; particularly when new orders cannot be released to the floor. Nothing brings adequate attention to a problem better than shutting down an operation or an entire plant until the problem gets resolved.

It Just Doesn't Seem Right – Or Does It?

There are some readers that cringe at the thought of a work stoppage. It is not natural to some to stop production. After all, aren't you in the business of making product, not sitting around? Why on earth would you want to stop working? Think of the ramifications: People would be sitting around doing nothing. On the surface, the arguments seem logical until you examine the alternatives.

If the problem being experienced downstream were caused by a quality defect that was created upstream, then continuing to deliver more parts would create even worse problems. A massive amount of scrap parts or a huge rework effort could be the result. Instead of being hours behind, the plant could find itself three or four days behind on some or all of its orders. Excessive WIP creates excessive work when problems crop up.

If the workflow were stopped, using the same quality scenario there would be a minimal amount of scrap and less rework to deal with. Once the problem is fixed, normal production could resume in a minimal amount of time. Stopping the flow of work would promote the focused attempt to determine the root cause source of the problem. In those areas upstream from the problem area, training could be conducted; PM's could be performed on machines and department meetings could be held, **or** the 'idle' employees could be used to help fix the problem. Stopping the flow of work need not result in people sitting around doing nothing, but even if that were the case, it would be better than allowing the big buildup of WIP. Dealing with excessive WIP only causes you to do a lot of Non-Value Add work.

Dynabans (Dynamic Kanbans)

While Kanbans are controlled queues with upper and lower limits, Dynabans are controlled queues with multiple buckets, each of which has its own upper and lower limit. Dynabans are used to control the

WIP levels for producers whose product line are large and contain many product complexities.

If you are lucky enough to have just a few different part numbers and they are all essentially similar when it comes to complexity, then Kanbans are perfect for you. But if your parts list numbers in the 10s, 20s or more, you may need Dynabans. The issue here is not that you have a lot of part numbers to deal with, the issue is that some of those part numbers are considerably more difficult to produce than what is called vanilla or standard parts

As companies take on new customers, some of those customers may have parts requirements that present some very difficult and unique challenges. In the printed circuit board production industry, for example, the complexities may involve multi-layer boards, trace separation, size, and shape or component population. Each of these issues can make the PCB's more difficult to produce because they require more production time than standard product.

Some product is so much more difficult to produce than "normal" product that if only those parts are run, the line would essentially come to a halt. In order to meet requirements to all of the customers, the difficult product should be "metered in" to the production floor in "digestible" quantities. Dynabans have proven to be very effective in controlling the amount of the difficult product that is in the line at any given time. The entire factory need not suffer from too much difficult product, which is another guaranteed outcome if product complexities are not properly identified and managed effectively.

In large production companies, certain part types only present difficulties to a few work centers, while having no effect on the majority of the line. When this situation is the case, Dynabans may not be needed throughout the plant. Even so, the status of those Dynabans must be known by those responsible for releasing product to the line. Releasing too much or too little of any part number could either cause the line to unnecessarily slow down or cause the company to miss shipments of the difficult product.

It may seem hard to believe, but workers prefer working on product that is easy to produce when they have their choice, especially if those workers have numbers requirements to meet. If these requirements, or quotas, do not differentiate between difficult and easy products, then scheduling problems are certain to occur. In such cases the workers prefer to work on product that will give them the highest output. The workers will quickly come to know a part that presents them difficulties. Unless there are controls in place that makes it impossible to "cherry pick", workers may have a tendency to avoid the more difficult product types, and concentrate on the vanilla parts.

To back up what I have just said, I want to remind you that Deming was very outspoken when it came to his opposition to numerical quotas for workers, and for a very good reason. Numerical quotas create an atmosphere for "cherry picking" because it places the highest priority upon numbers, not the proper mix of products. Once numerical output is a worker priority, it becomes very difficult to convince workers that they should spend their time working on products that present them with difficulties, because difficulties represent low numbers to workers.

Using Group Technology To Establish Part Types

Grouping the individual part numbers starts with the establishment of critical attributes. Critical attributes are those attributes that make those product types unique. Since we are only talking about **critical** attributes, there should only be a dozen or so critical attributes established. Using Electronic PWB, or panel manufacturing as an example, here are some of the critical attributes (examples are not accurate) that could be established:

1.   Odd Size – Greater than 24 X 36 inches
2.   Drill Holes – Greater than 10,000 holes
3.   Trace Separation – Less than .001 inches

The three critical attributes all present problems to manufacturing. Odd sized panels require special fixtures in some departments. The more drill holes required, the longer it takes to finish the drilling of the panels. When trace separation is less than standard, there are more defects found because of shorts and opens. Each of these attributes affects the lead-time of the product.

Once the critical attributes are determined, it is time to bounce each and every part number against the attributes. Those parts that cannot be associated with at least one critical attribute are considered standard, or vanilla, and would fall into their own technology group. An example of the attribute tagging process would go as follows:

Part Number	Critical Attribute		
	CA1	CA2	CA3
0000001	X		
0000002		X	X
0000003	X		
0000004			
0000005			
0000006		X	X
0000007		X	
0000008			X
0000009			
00000010			

Figure 6.1.1 – Tagging Attributes to Part Numbers

Looking at the Critical Attribute Matrix, Part numbers 0000004, 0000005, 0000009 and 00000010 have no critical attributes associated with them so we can put them into a vanilla group. Part numbers 0000001 and 0000003 are associated with CA1 so they will form the second technology group. Part numbers 0000002, 0000006, 0000007 and 0000008 could form group 3 in this example. The grouping could be as follows:

Product Group 1 – Vanilla
Product Group 2 – Large panels
Product Group 3 – Holes and Layers

The primary consideration here is that certain attributes affect certain departments adversely. Once the requirements are determined, a schedule is created for the release of product in quantities necessary to meet the lead times for each order. If the releases are late, the orders will be late unless they are prioritized. If they are prioritized, other orders will suffer. The preferable alternative is to release product with adequate consideration given to the cycle time of each product group.

It is much better to release a good mix of Group 1, 2 and 3 daily, rather than flood the floor with one particular group one day, another product group another day, and so on. When the tough product hits the floor in bunches, it will create a bottleneck in one or more departments. The vanilla will flow through like water through a sieve, but if the release of the vanilla is delayed because there is too much Group 3 on the floor, you have a dilemma. It is much more productive to meter in the tough product with the vanilla.

Dynabans were created to facilitate the metering in of tough product and the proper balancing of your group technology product mix. Each product group will have upper and lower limits. In a scenario where 70% of the product released is Group 1, 15% is Group 2 and 15% Group 3; the Dynaban limits will reflect those numbers. Of course, consideration is given to increased fallout with Group 2 and 3, but the total amount of WIP in the line should reflect those quantities.

As is the case with Kanbans, when the lower limits are reached, a signal goes out to the supplier to send more work. When the upper limits are reached, the supplier is signaled to stop sending work until the level is lowered, or better yet the work center gets some help to improve its output capability so that it is not out produced by the supplier (this is known as HELPING the gate, or Gate Busting). As is true with Kanbans, the situation of shutting down a producer could reach all the way to the initial release point, but only when every Dynaban is full. Each call for work is for a single bucket of the Dynaban. When the vanilla bucket reaches its lower limit, the call will go out for vanilla product only. The same rules naturally apply for maximum limits. When a bucket reaches its maximum, the stop delivery order is put out for only that individual bucket. Other product may be delivered to other buckets until they reach their limits.

Dynabans help to keep WIP levels low, as do Kanbans, however they affect individual part groups, or Technology Groups, individually. If a quality problem hits one particular Technology Group, it will not affect the flow of the other product groups unnecessarily. In this way, work delays are minimized because only the product group being affected will be delayed or stopped.

Chapter 2 – Kanban and Dynaban Applications

Naturally, the primary use of Kanbans and Dynabans is to control the amount of WIP that hits the floor, but they have other uses. They can be used to minimize the loss occurred during planned machine rebuilds or relocation. There are machines that require refurbishment that takes days to complete.

First, let's take a scenario where a critical production machine in Department E is scheduled for refurbishment. The job will take 5 to 7 days to complete. Since 90% of all product goes through this standalone machine, the company faces a very serious loss of output and late orders. The company does not want to lose customers to competitors so it has decided to continue to release product to the floor during the downtime and use the significant catch up capability of the machine to eventually make up the deficit.

There are a couple of problems with the strategy: First and foremost, there is a possibility that the WIP buildup during the downtime could have quality issues that will not be detected until the machine is brought back on line. Second, once Department E comes back on line, it has enough problems just trying to catch up; it does not need the added problem of trying to deal with quality issues. Once Department E is up and running, it should be able to focus on just getting back to normal production. Finally, the solution is not really a solution, because it does not minimize the loss, it simply accepts the loss and tries to make it up.

What is needed here is a solution that is designed to minimize or eliminate the loss of output for the time that the machine is down. It goes without saying that there should be an attempt to reduce the time that it takes to do the refurbishment. Setup reduction principles can help in this regard, but that alone is probably not enough. A complete refurbishment can be very complicated and could involve factory representatives who are not guided by setup reduction techniques. Concentrating on the refurbishment time may be part of the solution, but it is probably not enough.

Enter the **Strategic Kanban** or **Dynaban**. To start, a Strategic Kanban/Dynaban is described as a temporary Kanban/Dynaban that will be discontinued once the need for it is over. It is strategic in the sense that it meets a specific need that is out of the ordinary. In this case, it will be used to minimize output loss.

The Strategic Kanban in this case will be placed behind Department E and will contain finished product processed by Department E. The product will be placed in the Strategic Kanban little by little. Work will build up in the Strategic Kanban slowly and delivery to Department F will continue normally during the buildup in a First-In-First-Out (FIFO) basis. Work should not be stored early in the Kanban and just left there once it is full. It is important that the buildup be accomplished over time so that the product stored in the Strategic Kanban does not become aged. Since product will be processed in a FIFO manner during the buildup, the product will be relatively fresh when the downtime begins.

It is important to carefully calculate the amount of time that it will take to successfully fill the Strategic Kanban to its required size. Building up too much work would be counter-productive. The calculated size should be sufficient to continue shipping product from Department E during the five to seven day refurbishment effort. Production in Departments A through D should continue to produce product at a reduced level during the refurbishment. Production in Departments A through D should be reduced to about 20% of normal during the refurbishment in this case. By the end of the refurbishment, there should be only one to two days worth of work in the Kanban/Dynaban in front of Department E. From there, it is back to normal production.

The use of Strategic Kanbans and Dynabans can be used in other, similar situations, but the important thing to remember is that they should never become permanent, nor should they be used frivolously. My experience has shown that there are many production managers out there that feel better when they have some extra work around waiting to be worked on. It seems to be something like a security blanket for them. Without a bunch of WIP at the front, they just do not feel comfortable.

It is difficult for many managers to get used to the idea of having the minimum amount of work necessary to meet their requirements. Low WIP levels can be traumatic to some until they realize that it is so much better for their business. Consequently, many managers look for ways to increase the upper limits on their Kanbans. Some of them will even arbitrarily change the upper and/or lower limits on Kanbans or Dynabans for no apparent reason other than the fact that they perceive some sort of need to do so.

Always remember - more WIP does not mean more output for the factory! The output capability of a department or factory is what it is. We should always work to improve our output capability, but not by adding WIP to our line. WIP and Kanbans should be sized appropriately for the capability of the department and/or factory.

Chapter 3 – Implementing Kanban/Dynaban Management

Implementing Kanban/Dynaban Management need not be complicated, but it certainly should be well thought out and coordinated. Usually, there are a number of work centers where Kanbans will be established at the input and output locations. Add to that the fact that there could be additional Kanbans within the work centers. The discipline that is needed for Kanban control to ensure that the upper and lower limits are not violated must be accomplished through training. Both management and workers should be well trained. Workers that will be responsible for monitoring and filling the Kanbans should not only be trained, but also certified to ensure that they both understand the run rules governing Kanban Management, and will comply with those rules.

How Much Is Enough? How Much Is Too Much?

What should the upper and lower limits be for each department and work center? Is there an algorithm that can be used to determine the limits? The answers to those questions is "It depends" and "Yeah, in a way".

Each department will have upper and lower limits that are determined by examining the following three factors:
- Reliability of on-time delivery from the supplier.
- Desired X Factor
- Daily output **capability** of both suppliers and customers.

Many times, a work center feels uncomfortable about lowering their WIP level because they are concerned that their supplier will miss a shipment to them and subsequently, they will run out of work to do. In production, you do not want your people to run out of work and this is especially important to people who work in the constriction area(s). Remember, when a constriction misses output opportunity because of a lack of work, it cannot possibly make up the loss at normal cost.

The establishment of the lower limit is determined by the examination of two factors:

- The probability of missed deliveries by the supplier
- Output potential of the supplier (Catch-up capability)

TAKT Management can be used to determine just how often each department experiences problems resulting in a failure to deliver product to its customer. For example, Departments A and B are sequential operations, and Department A has a stand-alone machine that requires one hour of maintenance every day, but is otherwise very reliable. Department B will need one extra hour of work plus machine startup time in their Kanban, or they can expect to run out of work as a result of no WIP. If deliveries are made hourly, as they should be using TAKT Management, Department B should have a Kanban lower limit size of 2 hours of work.

In establishing lower limits, it is only necessary to examine the **probability** of a missed delivery. By examining the maintenance records, it will be easy to determine what the probability of a missed delivery will be because of machine downtime and what the effect that the missed delivery will have on production. To determine the facts, you can use process mapping and simulation modeling, along with any empirical data that has been collected including TAKT data and maintenance records.

When establishing limits for a Kanban/Dynaban implementation, the tendency will be for departments to add a little extra for both limits. This is understandable at the beginning and allowable because you certainly do not want to turn anyone against Kanbans, you want them to become comfortable with them and for them to learn to respect them. The limits that you set should be designed to present you with an X Factor of around 5, (Remember that X-Factor is the amount that we can multiply our Raw Process Time, RPT, by to get our Cycle Time, CT) and a very reasonable cycle time for your product. Setting the limits too low or too high could cause workers and management to lose confidence in Kanbans as effective tools.

The implementation period should be one where initial values are set a little bit loosely, but monitored constantly to determine whether those first values were proper. If it is determined that there is possible justification for modification, the authority that established the values should re-evaluate the limits periodically and make any necessary corrections to them using the same criteria that was originally used. The re-evaluations should not be conducted on whims, but on meaningful data. Remember, the situation on the line changes constantly, but changes should be made only after **sustained** justification.

As problems are fixed such as maintenance problems and delivery issues, the limits should be adjusted to reflect the improvements. The decision to adjust limits should never be the result of increased requirements. The way to deal with increased requirements is to increase capability, not to set Kanban limits. If the output capability and requirements are increased and a one-hour buffer is needed in the Kanban, then the one-hour amount should result in an increase of the upper limits.

Simulation modeling is a powerful way of establishing upper and lower limits. Simulation can give you the "big" picture by showing how the limits that are set affect the entire line, rather than just guessing what the effects of the changes will be. Dependencies of one department on the next and statistical fluctuation of a department's output capability can be effectively modeled using simulation. I have used simulation on numerous occasions to establish Kanban limits, and I still believe that it is a wonderful tool, but if you do not have the ability to use simulation, you can still implement Kanban/Dynaban (K&D) Management effectively using the methods outlined above.

The real power of K&D Management is realized when there is a continuous improvement policy in place. When TAKT Management is being used and constrictions are being identified and bottlenecks broken, then the job of K&D Management becomes much easier. But the real secret, if you call it that, to an effective implementation is **discipline!** If the limits are respected by management and workers, K&D management will effectively control WIP and help to establish and maintain a world class cycle time and X-Factor. If used in conjunction with other FCP tools, such as TAKT Management, Constriction Management and Process Mapping and Analysis, the probability for the attainment of World Class status increases exponentially.

K&D Management should **always** be implemented simultaneously across the entire line. Never try to start it here, then there. Plan and prepare for it carefully, then turn it on all at once when you are ready. Once the K&D locations are established and the limits set, the run rules governing K&D Management are no longer optional. Compliance becomes mandatory for everyone. Of course you want input from everyone, but you want that input prior to implementation. If there are concerns about K&D locations or limits, the time to debate those concerns is best accomplished prior to the start of the implementation. Once the line is turned on, then only the establishing authority can make changes.

It is best to use a Core Team to implement, train and coordinate K&D implementation. It is best when a central authority is responsible for the data research, coordination, setting of limits and the establishment of Run Rules. The Core Team can pull resources from individual work centers for help, and involve other departments or support areas, but the prime responsibility for putting it all together should be centralized. When management understands that the authority for making changes rests only with a team, rather than with individual managers or supervisors, the likelihood of violations is lessened considerably.

There should be Run Rules in place that place severe restrictions on overriding established K&D limits. The authority to override the established limits should always require the approval of upper management. The authority for overriding established limits should never be given to supervisors or middle managers. It should be extremely difficult, if not impossible for production managers to take it upon themselves to change the upper or lower limits of a Kanban or Dynaban.

Part 7 - Run Rules

Tom Clason

Chapter 1 – Run Rules Help To Maintain Process Integrity

Remember that Run Rules are not just written procedures. They do not just tell workers how to perform a technical task. Run Rules do not just cover the business as usual scenario, they should completely cover alternative scenarios. They are not used to evaluate personnel on how well that they do their jobs, and most important of all, they are just as applicable to management and support groups as they are to workers.

The establishment of Run Rules should begin with Management Run Rules. Management should make it clear when, and under what conditions, they would order any deviation from process Run Rules on the production floor. It should be crystal clear what the extent of their deviation would be from the existing Run Rules for any given situation. In other words, Management Run Rules should be a definitive statement saying how management will run the business under a variety of conditions. Management Run Rules should be written for executive management, production management and support management. By establishing and following their own set of Run Rules, management sets a positive example for all of the employees.

What would your company do in the following scenario: You have just suffered a very serious maintenance issue. One of your machines has broken down and you did not have the necessary part to fix it. The machine has been down for four days waiting for the part to arrive and be installed. You are up now, but you have lost a great deal of production time and you have a number of orders that are overdue. Customers are calling in and demanding their product. The calls from customers are coming in from angry and anxious customers, some of which are stating that their own businesses are in jeopardy if they do not get their product.

What should management do in this crisis? Should they respond to the customers with the most serious issues? What if they (management) push orders that would cause productivity to suffer? What if the orders that are considered the most critical would cause the line to come to a crawl (critical attributes) if those orders were pushed ahead of others? What is the proper course of action for management?

There is no textbook answer to the above questions other than the obvious: Management should do whatever is best for the majority of their customers while using Run Rules that are consistent with good production values. When I say that consideration should only be given to a small percentage of customers, I mean just that! The prioritization percentage should depend upon the effect that it (prioritization) had upon X-Factor. If the problem experienced has resulted in a large amount of WIP to accumulate on the floor, the X-Factor will be highly elevated. In such a case, the prioritization should be held to a minimum.

Remember, with an X-Factor of 5 or less, 50% of the orders could be prioritized without a serious risk. If the X-Factor were elevated to 10, 15 or more, attempting to prioritize orders would be extremely dangerous. In a situation like our scenario, the goal should be to minimize the negative effect that the problem created, not to increase the risk of additional problems.

Sticking to the main point here, there should be Management and process Run Rules in place that allow both management and production to respond **predictably** to this and other crisis. After all, that is what Run Rules are for – to allow a company to respond to unpredictable circumstances without chaos quickly and effectively.

Imagine a company without management and production Run Rules. If a customer called one manager, they might get their order prioritized at the expense of other orders. Another manager would react to a call from another customer, or if a customer failed to get a positive response from one manager, they might go to that manager's boss, or even the plant manager. I have even known of customers that went higher than the plant manager to get their order prioritized. When managerial pressure goes downhill, it is much more likely to get results, even if those results have a negative effect on production.

When there are Run Rules in place that prevent that managerial pressure from adversely affecting the production line's ability to process the orders efficiently, the company will be able to minimize the negative effect of major problems because the solutions to those problems have been thought out in advance. There need be no knee jerk reaction to unforeseen problems when good, value oriented Run Rules are in place.

Writing The Run Rules

As is true with Kanban/Dynaban sizing and Setup Reduction, the preferred method of establishing Run Rules is to use the team method. There is a lot of coordination necessary to get the Run Rules written. Some rules will require mostly management and their key production people, and others will be more technical, requiring engineering, quality, maintenance and worker involvement. The facilitator/coordinator will be the one that determines who is needed. Obviously, it is important that the facilitator be very knowledgeable in the process and in the proper production values and tools.

Once the proper people are assembled, the first session should begin with a review of the current procedures if there are any. If there are no formal procedures, as may be the case with procedures for shift changes, which may vary from shift to shift, process mapping and analysis will determine the different ways that the process is executed. It may well be necessary to **film** the present process to give the team an understanding of how the job is presently done. As in just about every productivity improvement effort, understanding the baseline is essential. You cannot determine where you are to go if you do not know where you are. Process mapping is an excellent way for the Run Rule Team to understand the baseline flow. In some situations, simulation modeling can be an outstanding way of both understanding the baseline and determining what the new process should be through "what-if" analysis.

The Run Rules should be written in chronological order when appropriate, but should certainly cover the rules under a variety of conditions. Unless there are a variety of "what-if" scenarios, the Run Rules are not completed. In the production business, you must anticipate the variety of possibilities that could create confusion if people do not know how to deal with them. Normal production issues do not normally inhibit productivity, but unforeseen occurrences certainly will. If the people that are hit with the unforeseen do not know how to respond, productivity can be lost, and sometimes the loss can be significant. Run Rules deal with the potential confusion in a positive and effective way.

Flowcharting is an excellent way of writing Run Rules. Run Rules that are flowcharted are easy to follow, and can be color-coded for even more clarity. If there is a question half way into the procedure, it is easy to determine just exactly where to look. Flowcharted Run Rules are easier to commit to memory because they are more visual. If Run Rules are printed, the font should be larger than normal. Never use small font when printing Run Rules. They should be easy to read and uncomplicated.

Although flowcharting is a very effective way of writing Run Rules, it is not absolutely necessary. Consider the example of a simple Run Rule (Figure 7.1.1) that deals with the discovery of a Kanban whose upper limits have been exceeded.

Temporary Run Rules for Mechanical Assembly
3rd shift only
Reason: Kanban Over Limit at Start of Third Shift

- **Try to contain WIP with current staff size first. Do not fold people from other departments: inspection, test and final inspection**
- **Fold people internally, if available (within Mechanical Build Area first, then other departments, if available) - To Label & Label/Build as needed**
- **Return "Folders" to their original departments once Kanbans return to normal levels**
- **Post current Kanban number on Takt Chart at 4AM**
- **Report Occurrence to Department Technician**
- **Technician to Present Alarm Tracking sheet at 7:30AM Takt meeting if Kanban remains over upper limit**

Figure 7.1.1

Figure 7.1.2 is another example of a Run Rule that deals with new product engineering builds that must be included within the normal production flow. One of the issues that this Run Rules deals with is to control the percentage of engineering builds that manufacturing must deal with at any given time. In this run rule, engineering is required to meter in the engineering builds so that normal production is not seriously hampered by an abnormal amount of engineering builds. If the amount of builds is coordinated, it will not be disruptive to manufacturing. This is a win-win Run Rule that was developed by Production Control, Engineering and Manufacturing.

Engineering Build Run Rules

- **Engineering Build's will require their designated Color Code Label**
- **Engineering Builds are not part of the Kanbans**
- **PC, Mfg., and Eng will commit to Priority/Build Schedules.**
- **Engineering Builds are "metered" out to Mfg. not to exceed 20% of DGR**
- **Completed Kits are Required Prior to Release**
- **I.E. Complete Kits and Build Hardware**
- **All Programs, Fixtures and Tooling are required prior to release**
- **Fully trained 3 Shift Mfg. personnel is required prior to conversion to production Order Color Code can be used (FIFO to be followed)**
- **All Documentation/Assembly Instructions must be on the floor prior to conversion to a production order type**
- **Record Engineering Build Quantities on the Takt Charts**
- **Exceptions must be approved by two levels of Mfg. Management above Supervisor**

Figure 7.1.2

Figure 7.1.3 is a partial example of Product Release Run Rules that specifically deal with issues related to the initial release of product to the line. In addition to some initial release rules, the Run Rules also address the issue of QTA's (Quick Turn Orders) that must be processed in less time than normal orders. Since QTA's receive priority on the line, they will affect the on-time delivery of other orders if too many QTA's are released during a shift.

Product Release Run Rules (2/19/99)

1. **No QTA will be scheduled for a cycle time that is under 80% of MFG CYCLE TIME**
2. **Max two QTA's (either crib) per day, Monday thru Friday. Limit 10 QTA's/wk**
3. **Only QTA's will have red folders**
4. **Re-releases due to composite fallout have the same queue jump capability as QTA's**
5. **There will be a limit of two product re-releases/day due to composite fallout.**
6. **A recommit on a re-release cannot be </= 80% CT**
7. **Quality issues must be resolved on the same shift**
8. **Those that cannot be resolved on the same shift must be put on HOLD - Refer to the On Hold Procedure**
9. **Product that has been designated as scrap can only be re-qualified by Quality   Engineering.**
10. **FIFO will be practiced in all areas except for the following conditions:**
- **A QTA can jump a maximum of three places in the input queue**
- **After a QTA jumps place, at least one FIFO lot must be run before another QTA is processed.**
- **Previous week's colors take priority in the input queue, and must be processed before the present week's product.**

Figure 7.1.3

Compliance – The Biggest Obstacle!

It is one thing for Run Rules to be written, and sometimes quite another thing for them to be respected by everyone in production. In some plants where Run Rules have been written, compliance is completely at the discretion of management. I have known of managers and supervisors that have walked straight up to a worker and order them to disregard the Run Rules. One example of this was a Run Rule that stated that once a setup was completed and an order was started on a machine, the machine would continue running until the last part in the order was completed. The reason for the Run Rule stemmed from the fact that tearing down an uncompleted order was wasteful and could (probably) affect quality. Still, the supervisor ordered the worker to tear down the order, set it aside and complete a setup on another order.

The existing Run Rule took into account the possibility of a situation where an order needed to be pushed ahead of other orders because of a executive business decision. The new order would be pushed ahead of all other orders in the input queue and would be processed as soon as the last part on the existing order was completed. Sadly, the supervisor's actions adversely affected the delivery dates of a number of orders and the order was not shipped a bit earlier than it would have been if the Run Rule had been respected.

The best course of action is to take great care in establishing Run Rules and then to make absolutely certain that they are respected. There are ways to make sure that the workers are following Run Rules. Managers can make it clear to the workers that compliance is mandatory. But unfortunately, managers and supervisors are the biggest violators of Run Rules. Usually, if a worker breaks a Run Rule, it is because his or her supervisor ordered them to do so. The most effective way of ensuring Run Rule compliance is for management to show leadership by setting a good example.

Although it is very important that the Run Rules be written carefully, it is equally important that the Run Rules be complied with. This problem cannot, in my opinion, be overstated. Non-compliance is a very insidious problem that can hurt production without giving a clue that it is the real reason for the loss. People do not shout out that they are not following Run Rules. Supervisors do not order the breaking of Run Rules and send a memo out to others admitting what they did. But when output drops and the result is lost business, it is almost too late to respond to the problem. You can correct the problem, but you cannot regain the loss.

You cannot take a chance that the Run Rules are being followed to the letter. You must establish and maintain no-notice audits to make sure that process discipline is maintained. Setup Reduction may open a bottleneck, but that bottleneck can reappear just as easily if vigilance is not maintained. WIP can build up very quickly at a pinch point if too much product is released or if the wrong type of product is released. One shift can run only easy product, leaving the difficult product for another shift to deal with. The offending shift may meet their "numbers" to the detriment of the complete factory. Offenses could be both intentional and non-intentional. In either case, they would hurt the business.

Occasional no-notice audits are a very effective way of ensuring that Run Rules are being followed. First of all, everyone wants to be seen as successful. If people know that a no-notice audit can hit them at any moment, they are more likely to make absolutely certain that they are following their Run Rules to the letter. The intent of no-notice audits is not to create fear, but to increase awareness of the importance of the Run Rules.

In a perfect production world, using no-notice audits to maintain compliance would be unnecessary. Obviously, there is no such thing as a perfect world, either in production or anywhere else. Audits are used because mistakes happen and audits are very good at catching the mistakes before they become a big problem. Failing an audit and correcting the situation is much more acceptable than allowing a problem to fester until it adversely affects the ability of the company to satisfy its customers.

Maintain Discipline But Do Not Stifle Creativity

In a dynamic and innovative company, the constant search for improvement certainly extends to Run Rules. Each Run Rule is created taking into account a number of factors. Depending upon the type of Run Rule, a change could not only affect output, it could affect product quality. For this reason, changes should be considered very carefully. A suggested change may look very promising on the surface, but could have harmful effects on output or even quality if implemented. For this reason, only the team that establishes the Run Rules should make changes to the Run Rules. Others in production may recognize opportunity and make suggestions, but only the establishing team should be authorized to implement the changes.

First of all, if suggestions are made, the team must investigate the suggestions to ensure that they actually have merit. Since quality is always a priority, if the suggestion is implemented, it should have a positive effect on quality. Although the proposed change could have a positive affect on the department that is recommending the change, it could have a negative effect on the bottleneck department, making that department even less productive than it is. For example, the suggestion could be to increase the batch size in order to improve the utilization percentage of a machine. The result would be better utilization of the machine, but if the constriction could not handle the increased load, the suggestion would have a negative impact upon the company.

One of the rules of constriction control is to never make rules that have an adverse affect upon the constriction. Finally, the change itself may not be based upon good production values, which is the case in this example. With FCP, we look for ways to reduce batch sizes, not increase them. In this case, the suggestion is simply not a good one and the Run Rule should not be changed. Remember, any Run Rule that is based on bad production values **will** be counter-productive.

If it is determined by the team that the suggestion has merit, there is more to do than to just document the change. There is training to conduct, not just on the literal changes, but also on the reasons for the change. Everyone that is affected by the change must be shown the logic for the change and understand what the net effect of the change is projected to be. After all, why would anyone want to change a Run Rule unless the change will benefit the process? The effect of the change would have to be positive or there would be no reason to even consider the suggestion. Once again, simulation modeling would be an excellent way of testing a solution to determine the net effect that the change will have on the process.

Tom Clason

The Product Release Point

The perfect place for good Run Rules is where the product is first released to production. There are so many things that can happen, both on the production side and the scheduling side. Schedules are modified, cancelled or even put on hold. At times, an order that has already been released may run into problems in production, bringing up the need for more product to be released to make up the deficit in the order. The possibilities of issues arising at the product release point are endless. Good Run Rules at the release point could save the company a great deal of problems and profit.

What is the person at the release point to do if a supervisor asks that the schedule be altered? What happens when an order is to be released and a certain part is unavailable? What should happen if the priority of an order is re-classified? Run Rules would tell the worker exactly what they should do in each case. The worker would not have to call their supervisor or manager and ask them what to do if the Run Rules were written properly. They would know what their manager would say because the team that wrote the Run Rules included management. Management signed off on the Run Rules. There should be no hesitation on the part of the worker because the training included the worker and management. The worker would respond to each issue quickly and decisively.

In the case of the supervisor requesting that the schedule be altered, the Run Rule could state that the schedule will not be altered unless there are two levels of management signing off on the change. The supervisor would have to get his or her manager to agree to the altered schedule. Of course, a Management Run Rule may be written that disallowed the supervisor from interfering with the schedule. In that case, the supervisor would be breaking a Run Rule just by asking for a schedule change. But the worker would know exactly what to do and how to respond to the change without jeopardizing their position with management. The only damage that the worker could do to himself or herself would be to violate their Run Rule.

Part 8 - Organizational Cultural Improvement

Chapter 1 – The Birth of The Combined Offering

I have known for some time that I had a good system for improving productivity. My system worked every single time and never failed to get great results. I knew that I had a good product, but I also knew that there would be resistance to change in each and every project. There were times when it was very difficult to get the new system implemented – not because the methodology was flawed, but because there was resistance to change that appeared at inopportune times. Sometimes the resistance came from very formidable sources and sometimes it was very difficult to identify the source of the resistance. I knew that the most likely source of resistance would come from production supervisors and support people. The least likely source of resistance would come from workers. In most cases, the source of the resistance was very difficult to identify.

Being a bit of a bulldog by nature, I would do everything in my power to convince the people at the top that I needed the cooperation of the managers, supervisors and technical leads in order to get my job done prior to the start of a project. Once I had their assurance that they would support the effort, I trained the Core Team. I taught them good production values and how to implement and use the FCP tools that would help their company to be successful. I turned each Core Team Member into an internal consultant and an expert in FCP. Once they were trained and ready, the Core Team would then analyze the process prior to the implementation of FCP.

Once the team finished their analysis and initiated the implementation process, everything went very well at first. The Core Team would be highly motivated and would work very hard to do what they believed would make their company successful. In order to implement the new system, the Core Team needed the cooperation of both management and workers. In addition, there would be new Run Rules that would affect the entire company, including management, engineering and maintenance. Of course, we would need their cooperation also. At times, the necessary cooperation was slow to arrive. Occasionally, there was open resistance from some members of management and support personnel. Occasionally the resistance would be passive. We would be told that our efforts would be supported, only to receive no support at all. Even though I tried to prepare the Core Team for this resistance, it always seemed to surprise and disappoint them.

The resistance never completely stopped the implementation effort, but quite often it slowed the process down. In several projects, the implementation effort was delayed for weeks because of internal resistance to the changes that the team was advocating. The delays sometimes had adversely affected the momentum and the motivation of the Core Team members. They could not understand why anybody would resist positive change that would improve the capability of their company to be more successful.

I was shocked and surprised the first time that I received criticism when changes were implemented that improved the output capability of a company that had been working a great deal of overtime. I was shocked because the criticism came from both supervisors and workers. More than a few people told me that they considered the overtime pay part of their salary and that they depended upon that money to pay bills. I would not have been shocked if the amount of overtime was reasonable, but in some cases, both management and workers were on a seven-day workweek. The supervisor's reluctance to see overtime reduced really surprised me because supervisors do not get extra pay for overtime. What I found out later was that the supervisors were sometimes compensated for their overtime with bonuses and time off during the week. Still, I could not help but wonder why anyone would prefer such a demanding work schedule.

I would later discover that the reasons for the resistance varied. Some employees may have become reliant upon the overtime pay and just did not want it to go away. Others had different reasons – some of which had nothing to do with a desire to work more overtime. Some people resisted change, even though they believed that the alternatives would be an improvement. After all, we were asking the people to change the way that they are working **and** thinking. We were asking the people to change procedures that they themselves initiated. We were asking them to accept ideas that they did not invent.

In some cases we received very strong, vocal resistance to the changes that we advocated. Now, I don't want to give the impression that the resistance was highly organized or that the workers that resisted change were in the majority. Actually, the opposite was true. In each and every case, the open resistance came from only a few people. But open resistance is the easiest type of resistance to deal with because it came from people that felt strongly about their opinions. In cases of open resistance, once we were able to convince the critics that we were right, rather than that they were wrong, we found that the critics became our best advocates.

Passive Resistance – The Hidden Enemy

The most difficult people to deal with were the ones who were unconvinced and would not let you know it. Their type of resistance was the most insidious because you could never be certain just where it was. If there were changes in place that didn't seem to be working as well as they could, how could we be sure just why? If the information on the TAKT Charts didn't seem to make sense, we could only assume that there was a need for additional training. For years, I struggled with this problem with varying degrees of success. I needed exuberant cooperation, not half-hearted semi-cooperation and I just could not figure out how to get it.

Let me tell you that when passive resistance hits, it is very difficult to work through it because there just aren't any symptoms to work with. Occasionally we will catch someone who just refuses to follow Run Rules, violates Kanban or Dynaban levels or just decides to perform a setup "their way." When confronted with the proof of their non-compliance, some of them will not even deny the fact that they are not following the rules. They explain themselves by saying that they didn't really understand them.

Power To The People!!

Another very serious form of resistance came from power-oriented individuals in responsible positions. There are people that believe that power rules. The power oriented supervisor or manager can make a productivity effort very difficult. They want to be seen as being in charge at all times and do not enjoy having their turf challenged. They want to be briefed before anything can change in their areas and they often hold up changes that would improve productivity in their areas until they decide that it is time. People who are power oriented are difficult to deal with because they do not often respond to logic. They cannot be convinced because they do not want to be convinced.

Occasionally, we ran into people who worked in foreman (non-management) type jobs that fit the description of a power oriented individual. These technical and non-technical leads were promoted because they were the most talented workers in their department. Many of them handed out tasks to workers in their department and functioned essentially as supervisors without the authority that a supervisor has. Still, even though they could not hire or fire people, they had a great deal of influence with the workers in their departments. Some of the strongest resistance to change came from these worker leads in a productivity improvement project.

For many years, I was at a loss to deal with resistance that I received when I worked a project. First of all, I was always busy working with the Core Team. I had no time to go out there and try to find the people that were holding us back with their resistance. Even if I knew who the person was that was hampering our efforts, there was very little that I could do about it. I could only press onward and do my best to work through it. The enigma continued until I met Gerry Clarke.

We were each consultants working with the same client when we met. He was helping the customer to become ISO certified and I was working to improve productivity on their production line. Later, we would form a partnership that would last several years and three projects. Clarke introduced me to some powerful cultural improvement tools that were the product of a company named Human Synergistics. I was interested from the beginning and, eventually, I even attended a training class to learn how to use the tools. After looking into these behavioral modification tools, I came to the conclusion that it was possible that they would be very helpful to me in my line of work.

Would it be possible, I wondered, to remove barriers to change by using the Human Synergistic tools? Little did I know at the time that it would not only be possible, it would be proven on our very first project. Theoretically, it was supposed to take five years or more to realize significant cultural change in a company. What we discovered was that cultural change could be realized much more quickly if there was a catalyst. That catalyst for changing destructive behavioral patterns turned out to be **success**. When the HS tools were accompanied by productive success, the results were very dramatic.

The HS tools were very effective in locating cultural hot spots. The tools helped us to understand what kinds of behavioral patterns we were dealing with. Not only could we help ourselves by identifying and breaking down barriers, we were able to help the individuals in the company by helping them to change their behavioral tendencies from destructive to healthy. We utilized trained counselors that helped workers, leads, supervisors and managers to understand and deal with their destructive behavioral patterns and to develop healthy patterns. We were actually able to make a very positive change in cultural behavior patterns within the six months that we were engaged in the productivity effort. The five-year theory had been shattered.

The productivity effort turned out to be the most successful of my career. We didn't just improve their productivity; we improved their behavioral patterns and their corporate culture. People that would have been difficult to deal with were cooperative because we worked with them early in the project. We knew this to be true because the only resistance that we received were from support groups that were not included in the HS study because of project budget restrictions. We had tried very hard to convince our client that all support personnel and managers should be included in the effort, but were unsuccessful. The resistance came overwhelmingly from people that were not included in the cultural enhancement assistance.

The exclusion of the support groups from the cultural improvement effort, and the fact that the primary resistance to our efforts came from those groups convinced me of the power of the HS tools. Wherever we used the tools, the results were outstanding; where we did not use them, the need for them was very obvious. I became convinced that cultural improvement combined with a productivity improvement project in a seamless package would improve the results of the productivity effort dramatically. I also believe that the cultural enhancement effort was more than cost effective. It has proved its worth on each and every one of the projects where it was included.

Chapter 2 - The Human Synergistics' Tools

If I were to attempt to accurately explain the HS tools and the reasoning behind their creation, I would probably leave out some very significant information. From my perspective, they worked, and were powerful in their application. In order to bring true understanding to these wonderful tools, I turned to my partner in those very significant projects, Gerry Clarke, who kindly agreed to not only explain the HS tools, but to explain why they were so effective and how they worked.

From The Pen of Gerry Clarke ...

Fast Cycle Production, when implemented fully with strong commitment, is an extremely powerful methodology for process and system change and will always present a substantial return on investment for the implementer. However, as with all things, time can erode and change even the strongest of forces. The science of Chaos reigns: over time, small unnoticed changes in the systems, processes and procedures installed with FCP can produce large, significant changes to the overall outcome. Sometimes these are positive and desired; much more frequently they are not positive and not desired. Recidivism sets in and the entire process returns to a previous state, often worse than before the FCP implementation. These kinds of changes are usually extremely subtle, are hard to notice and almost always come as a result of the culture and climate of the organization.

In our work together, Tom and I found that combining FCP methodology with a full scale cultural intervention produced lasting results in the systems and processes changed through FCP and created much more healthy, constructive and effective cultures. This, in turn, created more healthy, constructive and effective leaders, managers and employees. Culture is the "thing" that holds it together. In the ancient Greek culture, all things consisted of fire, earth, wind and water. But there was an unseen fifth element that held everything together – ether. Culture is like that. It can be healthy, constructive and effective or it can be destructive, unhealthy and ineffective. But it cannot be "seen." It can, however, be measured.

Many projects proved these statements, but a particular project with a Fortune 50 manufacturing organization proved them beyond all doubt. For several years prior to our intervention, one of the company's microelectronics manufacturing facilities, which manufactured extremely complex computer panels, was doing very poorly. They were losing millions of dollars per year. Their 2500 employees were suffering from long tedious workloads, high overtime, few vacation days, and high stress. They were extremely behind schedule with shipments; their cycle time was horrendously long which was causing them to be uncompetitive in the marketplace; they were losing business to Far East sourcing; quality was poor, rejects were high and work in progress was stacked all over waiting for the next step in the manufacturing process. They were under extreme pressure from headquarters to improve their financial situation, but headquarters provided little or no help. In fact, headquarters wanted to sell the complex but could find no buyers for such a poorly performing unit.

Over a two-year period we introduced FCP and cultural change simultaneously, and the company achieved the following results:

- Overall cycle time reduced **68%** — 34 to 10.8 days
- WIP reduced **38%** — $21M to $13M
- Non-Value Add Operations reduced by $2.2M
- Behind-Schedule orders reduced 95.6% - 2100 to 91
- FTEs (full time equivalent employees) reduced by 27 ($600K)
- Overall inventory reduced $8M
- Capacity = 20% gain
- Culture — significant constructive shift in management
- 1 holiday weekend worked (vs. all for 3 yrs. previous)
- Moved from **$11M annual loss to $24M profit**
- Investment **$1M**

This company continued the emphasis on FCP and developing a constructive culture for four years showing growing profits each year. After the fourth year, a group of independent individuals bought the company.

Tom has explained the details of FCP; let's now cover the **cultural** side of the intervention.

Culture is most commonly defined as the basic *assumptions*, *shared values* and *beliefs* that *guide* the way organizational members *behave* toward each other and *approach* their work.

The Circumplex™

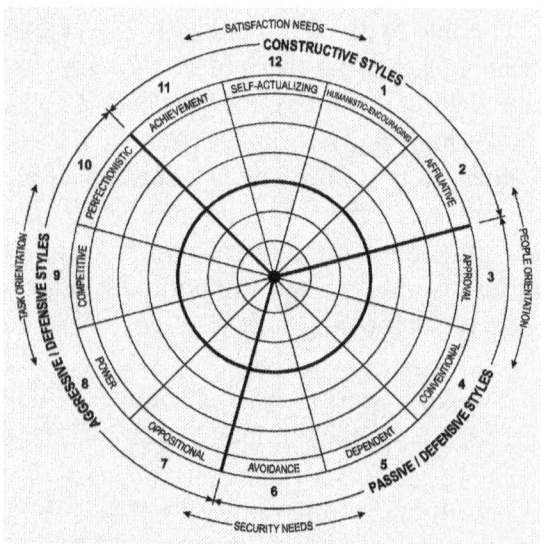

Figure 8.2.1 The Circumplex™

The Circumplex™ is the driving force behind a series of **Human Synergistics Inc. (HSI)** assessments: specialized surveys that measure the quality of performance at different levels throughout the organization. The series includes:

Organizational Culture Inventory™, or **OCI** for an organization's culture.
Life Styles Inventory©, or **LSI** for individuals,
Leadership/Impact®, or **L/I** for leaders,
Group Styles Inventory©, or **GSI** for groups/work teams

Linked together by the Circumplex™ and the 12 styles, these assessments form the HSI Performance System: a structured way of taking a close look at an organization to decide what's working, what's not, and how to fix it.

The HSI Performance Improvement System is used to diagnose performance difficulties and target the search for solutions. You can use the system to:

- Find out how well an organization is performing at different levels - individual, leader, work team, and overall.
- Combine this feedback to identify areas of concern - those "weak spots" that, over time, can send performance into a downward spiral.
- Decide how an organization could, and should perform. What should people be doing to take the organization where it needs to go?
- Determine the styles – specific patterns of behavior – that hold the organization back. And decide which styles people need to develop or strengthen to move forward.
- Give people the information and tools they need to make targeted, consistent changes to their behavior. Changes that, when combined, drive the organization to meet, or even exceed, its potential.
- Chart and follow an organization's progress toward improvements over time. Reassess performance and redirect efforts as needed to stay on the right track.

The Circumplex™ measures 12 styles of thinking, behaving, and interacting. The styles are divided into three *groups*, each with its own set of defining *characteristics*. Depending on whether they are Constructive or Defensive, these characteristics account for various types of results – at the individual, group, and organizational level.

Constructive styles reflect a healthy balance of people and task-related concerns and promote the fulfillment of higher order needs. Styles associated with this orientation are directed toward the attainment of organizational goals through the development of people. Constructive styles account for synergy and explain why certain individuals, groups, and organizations are particularly effective in terms of performance, growth, and work quality.

Passive/Defensive styles represent an unduly strong orientation toward people as opposed to tasks, fueled by and reinforcing individual insecurity. These styles characterize people who subordinate themselves to the organization but in the process, end up creating stress for themselves and allowing the organization to stagnate. Passive/Defensive styles can produce a predictable and secure situation, but at the cost of learning, adaptability, and ultimately survival.

Aggressive/Defensive styles emphasize tasks over people and are driven by underlying insecurities. In the extreme, these styles lead people to focus on their own needs at the expense of those of the group. Though sometimes temporarily effective, Aggressive/Defensive styles may lead to stress, decisions based on status rather than expertise, and conflict rather than collaboration.

In our projects, we measured the cultural norms and expectations using instruments from Human Synergistics and then began a program of working with all company leadership from the plant manager to the lead technicians. What follows are descriptions of the instruments used.

The Organizational Culture Inventory™ (OCI)

The *Organizational Culture Inventory™ (OCI)* provides a picture of the operating culture of an organization in terms of "how things are done around here" and "what's expected" of members. The behavioral norms and expectations measured by the OCI have a direct bearing on the functioning of an organization and, therefore, are related to specific *outcomes*. These outcomes include:
- Member satisfaction and motivation
- Teamwork and coordination
- Criteria or organizational effectiveness (such as quality of service and sales performance)

With the OCI, you can:
- Define an organization's current culture
- Create a vision of a more effective culture

- Identify changes necessary to move the organization toward that ideal

OCI results promote change by revealing the defensive and counterproductive behaviors the organization is inadvertently reinforcing, as well as the negative impact of those behaviors on individual and organizational performance.

Respondents can complete the OCI via the Internet, or by using the paper-and-pencil format in either group or individual settings. Through a set of 120 statements, the OCI measures 12 interrelated thinking and behavioral styles that members might be expected to adopt in carrying out their work and interacting with others.

Respondent scores are combined, and the composite scores along each style are plotted on the Circumplex™ to create a "picture" or visual profile of the organization's current operating culture. You can also request departmental, team, or other subgroup profiles to identify subcultures (and possibly counter-cultures) within the organization.

The OCI can also be used to create a vision of the organization's ideal culture by having members respond to a parallel form of the survey that measures what *should* be expected. Members are asked to identify the norms and behavioral styles that would promote their own motivation and performance, maximize the contributions of diverse subgroups, and help reach organizational goals.

The OCI provides a point-in-time picture of the culture of an organization in terms of 12 specific types of behavioral norms. It focuses on the behavioral patterns that members believe are required to "fit in" and "meet expectations" within their organization. These 12 types of norms are organized into three general clusters that distinguish between:

- **Constructive Cultures,** in which members are encouraged to interact with *people* and approach *tasks* in ways that will help them to meet their higher-order *satisfaction* needs (includes Achievement, Self-Actualizing, Humanistic-Encouraging, and Affiliative norms).
- **Passive/Defensive Cultures,** in which members believe they must interact with *people* in ways that will not threaten their

own *security* (includes Approval, Conventional, Dependent, and Avoidance norms).

- **Aggressive/Defensive Cultures,** in which members are expected to approach *tasks* in forceful ways to protect their status and *security* (includes Oppositional, Power, Competitive, and Perfectionistic norms).

In addition to measuring shared behavioral norms, the OCI also can be used to identify the ideal operating culture of an organization. A special form of the survey, the *OCI-Ideal,* measures the behaviors that members believe *should be* expected to maximize individual performance and the long-term effectiveness of the organization. The ideal norms can then be profiled against the actual or current norms to identify "cultural gaps."

The OCI form used to measure the *Current Culture* includes supplementary items designed to assess some of the outcomes associated with an organization's culture. Results along these items provide insights into the possible need for cultural change.

The OCI form used to measure the *Ideal Culture* includes supplementary items designed to assess members' (and the organization's) readiness for moving the current culture toward the ideal. Results along these items will identify potential sources of support as well as resistance to cultural change efforts and highlight additional effects that members anticipate will result from successful cultural change within the organization.

Promoting and maintaining a culture that encourages effective behavior on the part of members requires clearly and consistently communicating expectations across all organizational levels.

The 12 cultural norms measured by the OCI are arranged on a *circumplex™,* or circular graph. The circumplex™ is divided into three general *clusters*, each containing four of the 12 cultural norms measured by the OCI. The cultural norms contained within each cluster are similar to, or *work with*, one another.

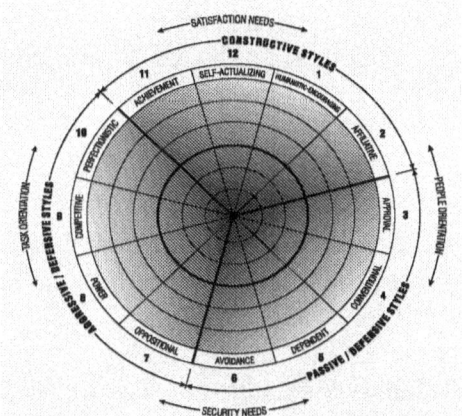

Figure 8.2.2 – The Circumplex™ Clusters

The cultural norms at the **top** of the circumplex™ promote the *satisfaction needs* of members and behaviors that enable them to fulfill those needs (e.g., needs for achievement and affiliation). The cultural norms toward the **bottom** promote *security needs,* and require self-protective behaviors associated with those needs (e.g., acceptance, avoiding failure). The cultural norms on the **right** side of the circumplex™ promote expectations for *people-oriented* behaviors; those on the **left** side are more *task-oriented.*

Based on the satisfaction/security and people/task distinctions, an organization can be analyzed in terms of three general types of cultures: Constructive, Passive/Defensive, and Aggressive/Defensive.

The Constructive cluster consists of cultural norms that promote the higher-order *satisfaction* needs of members. Constructive cultural norms include:

- **Achievement (11 o'clock)**
  Members of organizations with an Achievement culture are expected to set challenging but realistic goals, establish plans to reach those goals, and pursue them with enthusiasm.

- **Self-Actualizing (12 o'clock)**
  Members of organizations with a Self-Actualizing culture are expected to enjoy their work, develop themselves, and take on new and interesting activities.

- **Humanistic-Encouraging (1 o'clock)**
  Members of organizations with a Humanistic-Encouraging culture are expected to be supportive, constructive, and open to influence in their dealings with one another.

- **Affiliative (2 o'clock)**
  Members of organizations with an Affiliative culture are expected to be friendly, cooperative, and sensitive to the satisfaction of their work group.

The **Passive/Defensive** cluster consists of cultural norms that promote *self-protective* behavior in interactions with *people.* Passive/Defensive cultural norms include:

- **Approval (3 o'clock)**
  Members of organizations with an Approval culture are expected to agree with, gain the approval of, and be liked by others.

- **Conventional (4 o'clock)**
  Members of organizations with a Conventional culture are expected to conform, follow the rules, and make a good impression.

- **Dependent (5 o'clock)**
  Members of organizations with a Dependent culture are expected to do what they're told and clear all decisions with superiors.

- **Avoidance (6 o'clock)**
  Members of organizations with an Avoidance culture are expected to shift responsibilities to others and avoid any possibility for being blamed for a mistake.

The **Aggressive/Defensive** cluster consists of cultural norms that promote *self-protective* behavior with respect to the way that members approach *tasks.* Aggressive/Defensive cultural norms include:

- **Oppositional (7 o'clock)**
  Members of organizations with an Oppositional culture are expected to be critical, oppose the ideas of others, and make safe (but ineffectual) decisions.

- **Power (8 o'clock)**
  Members of organizations with a Power culture are expected to take charge, control subordinates, and yield to the demands of superiors.

- **Competitive (9 o'clock)**
  Members of organizations with a Competitive culture are expected to operate in a "win-lose" framework and work against (rather than with) their peers.

- **Perfectionistic (10 o'clock)**
  Members of organizations with a Perfectionistic culture are expected to avoid mistakes, keep track of everything, and work long hours to attain narrowly defined objectives.

We found that there was a great deal of difference between the cultural norms that we found at the manufacturing company that we worked with together, and the "ideal" culture that the employees there envisioned. The cultural assessment profiles and interpretations for the manufacturing company we worked with are shown in Figure 8.2.3:

Culture Assessment: May 1998

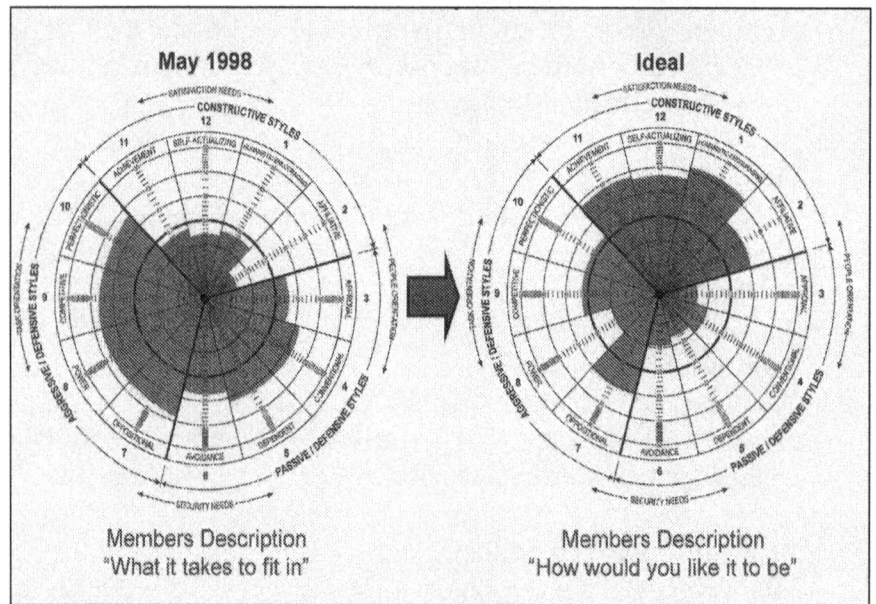

Figure 8.2.3

Organization Culture Style:
May '98
Primary Style: Aggressive / Defensive
 * Confrontation Prevails
 * Focus on what is wrong, not right
 * Management Not Involved
 * Reactive, Bureaucratic
 * Command and Control
 * "Win-Lose" Mentality
 * Work Long hours, Perfectionistic

Result

 * Roles Not Clear
 * Key Personnel Turnover
 * No Risk Taking
 * High level of Dissatisfaction
 * Lack of Customer Focus
 * Poor Financial Performance

Organization Culture Style:
Ideal
Primary Style: Constructive
 * Goal Setting and Achievement
 * Individual Growth
 * Value Creativity
 * Cooperation
 * Conflicts resolved constructively
 * Focus on what is right
 * Values Mentoring and Coaching

Result

 * Job Satisfaction
 * People "Go the Extra Mile"
 * Cooperative atmosphere
 * High Quality Customer Service
 * High Customer Satisfaction
 * Change is favored
 * Financial Growth Probable

As you can see in Figure 8.2.4, the comparison of the culture as it existed when we arrived and approximately a year and a half later showed a dramatic improvement. Our efforts to improve both the productivity and the cultural health of the company were successful on both counts. The results clearly showed improvements in the healthy, people oriented styles, and equally dramatic reductions in the destructive styles.

Culture Assessment
Comparison: May 1998 – December 1999

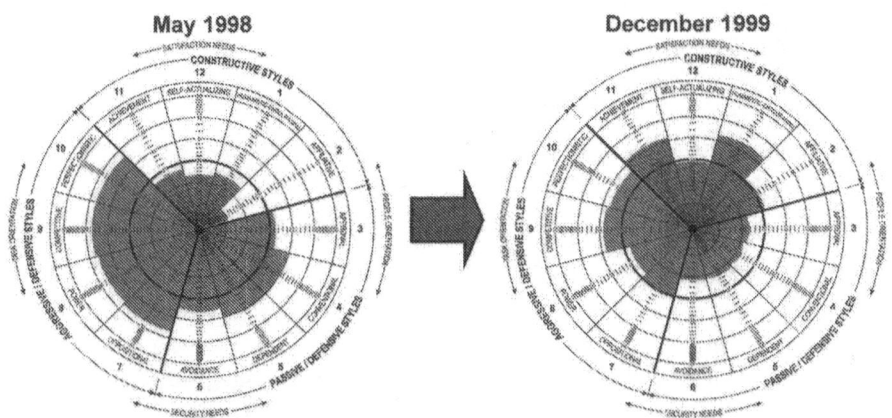

Figure 8.2.4 – Cultural Improvements

The *Organizational Culture Inventory™(OCI)* was used to measure the intangible aspects of the culture – behavioral norms and expectations. The *Organizational Effectiveness Inventory™ (OEI)* was used to measure and identify outcomes and levers for change resulting from the cultural norms.

The Organizational Effectiveness Inventory (OEI)

The *Organizational Effectiveness Inventory™ (OEI)* measures both *outcomes* (at member, group, and organizational levels) and the various *factors* (i.e., change "levers") that determine those outcomes.

Used in place of more traditional employee satisfaction or attitude surveys, the OEI improves on the information such surveys provide by indicating whether the organization needs to change, and pointing out areas for improvement.

The outcomes measured by the OEI are:
- Quality of Service
- Coordination and Adaptability
- Employee Outcomes-Negative
- *Person/role conflict, job insecurity, stress*
- Employee Outcomes-Positive
- *Role clarity, motivation, satisfaction, intention to stay*

The change "levers" measured by the OEI are:
- Philosophy and Values
- Quality of Communication
- Human Resource Management
- Appraisal and Reinforcement
- Influence
- Supervisory/Managerial Leadership
- Supervisory/Managerial Sources of Power
- Job Design
- Goals

You can use the OEI results to guide strategic planning, organizational design and development interventions, training and development programs, and human resource management initiatives. The OEI enables you to:
- Identify outcomes that indicate areas of concern
- Determine how these outcomes affect organizational performance
- Link positive change "levers" to the identified outcomes
- Decide which change recommendations to implement

Organization members respond to a set of questions regarding their organization, department, supervisor/manager, and job. Responses to the OEI are then combined, averaged, and analyzed. The organization receives a customized feedback report that presents overall and subgroup results. The report compares the organization's results to historical averages (based on up to 700 organizational units) as well as "Constructive Benchmarks" (based on up to 120 organizations with healthy cultures).

The Life Styles Inventory© (LSI 1 and LSI 2)

The *Life Styles Inventory© (LSI 1)* is a statistically validated assessment that enables an individual to take a close, revealing look at themselves for insights that direct and drive performance improvements. The LSI is normed based on the responses of over 9,000 individuals from organizations of varying sizes and descriptions.

In 20-30 minutes, a person completes the LSI by responding to phrases in terms of the extent to which they describe him or her. These phrases measure 12 unique patterns or "styles" of thinking and behavior people demonstrate as they complete tasks and interact with co-workers.

By transferring LSI scores to the Circumplex™, the person creates a profile or "picture" of his or her styles of thinking and behavior. These styles can be **Constructive** and support effective performance, or **Defensive** and prevent the person from working to his or her potential.

The LSI 2 – completed by five trusted associates – shows how others perceive an individual in terms of the same 240 items measured by the LSI 1.

By comparing **LSI 1** results with **LSI 2** results, individuals can identify gaps between how they see themselves and how others see them. These gaps point to areas of development toward which individuals can direct their improvement efforts.

The Leadership/Impact® (L/I)

The *Leadership/Impact® (L/I)* assesses a leader's impact in terms of the Circumplex™. Thus, a leader can motivate or drive others to behave in Constructive, Passive/Defensive, or Aggressive/Defensive ways.

To target areas needing improvement, L/I enables the leader to determine the impact he or she ideally would like to have on others. The Circumplex™ is used to highlight areas of difference between one's current and desired impact. By comparing the "current" to the "ideal" impact profile, the leader can identify opportunities for change that will move his or her impact toward the "ideal." This involves increasing **Prescriptive** strategies (which lead to a Constructive impact) and decreasing **Restrictive** strategies.

The Group Styles Inventory© (GSI)

The *Group Styles Inventory© (GSI)* enables teams to take a targeted look at what's contributing to, or detracting from, their success. It does this by capturing and "feeding back" to members their perceptions of the team's interaction styles. Combining and examining these perceptions reveals opportunities for change and improvement in terms of both what the team does and how it does it.

Following a problem solving session, each group member assesses the experience by responding to a set of 72 brief questions. Their responses reflect individual perceptions of the patterns of behavior or "styles" members exhibited as the team interacted to solve problems, make decisions, or set plans and policies.

The responses of each group member can be combined and plotted on the Circumplex™ to develop a composite profile or "picture" of the team's interaction styles. These styles can be **Constructive** and support effective results, or **Defensive** and prevent the team from realizing its potential.

By providing the team with targeted, easily understood feedback on its current state, the GSI results serve as a *needs assessment* that clarifies for team members what they do well, and alerts them to what they can do better – to maximize results and achieve synergy.

Implementing The Combined Offering

Each of the HS tools previously explained were used in each and every project that Clarke and I worked on together. To this day, I steadfastly insist that cultural improvement be included in all FCP projects. I am a strong believer of the effectiveness of these tools for very good reasons. As Clarke explained earlier, the results that we achieved, both in the areas of cultural enhancement and productivity were nothing less than astronomical.

Clarke and I proved that the simultaneous combination of productivity and cultural improvement was the perfect solution. The productivity improvements that were achieved were more than significant and the cultural changes were equally dramatic. What surprised us both was the fact that the cultural changes were achieved in much less time than was normal. We were able to accomplish in one year what would normally take five years or more.

Each offering acted as a catalyst for improvement for the other. The cultural change was achieved because the productivity improvement made people feel better about their jobs and their sense of accomplishment was improved. The productivity improvement effort was successful because the cultural improvements made people more amenable to change. The workers began to believe that their efforts were meaningful and management was actually listening to them. Management came to understand that true leadership begins with a concern for the mental well being of the employees.

After our successful projects together, I came to several conclusions. First, I concluded that it was absolutely essential to include cultural enhancement into every FCP project. Second, and what I consider most important of all, that the cultural team should begin their work **in advance** of the beginning of the productivity effort.

There are several reasons for the cultural study to begin prior to the beginning of the project: First and foremost is to determine the state of the culture and the location of any cultural "hot spots" in the company that could inhibit the total effort. In my opinion, a pre-emptive attempt to isolate those hot spots and improve the behavioral patterns of those that might offer resistance to our productivity efforts is essential to success.

In each of our projects there was no pre-emptive cultural enhancement effort. In one project, the cultural enhancement effort was delayed for a short period because the cultural enhancement team was still engaged elsewhere. In my opinion, the delay was costly because there was resistance to our efforts that would not have been there had the cultural team been in place early on.

In productivity improvement projects, timing is critical. Changing behavioral patterns deals with different issues than implementing a new Values Based production system. But there is also a common thread that unites each side of the combined offering. Each team is dedicated to IMPROVEMENT. Together, they provide a holistic cure

303

to the demons that keep a company from achieving world-class dominance.

There You Have It ...

It is my sincere hope that the readers of this book have by now developed an understanding and appreciation of the FCP values and tools. Obviously, FCP is not rocket science, but on the other hand, FCP is powerful when applied to an ailing process or a process that could use improvement without adding resources. If there is a need for improved output without increased spending, Fast Cycle Production is the appropriate solution. If, by now you believe this to be true, I have accomplished my mission.

EPILOGUE

By Michael V. Testani, Sr.

I truly hope that everyone who has read this book will recognize the power in Fast Cycle Production and its time-tested techniques. The real power of these productivity improvement techniques originates from their deeply rooted values. Fast Cycle Production's values-based manufacturing system is unrivaled by other, more popular, productivity offerings in that FCP addresses the fundamental principles of good manufacturing practices. A building that is built to last several lifetimes is built on a solid foundation. The same holds true for a Manufacturing company that wants to last for several lifetimes. It must base its entire existence on a strong, solid foundation upon which the rest of the organization is built. That foundation is the company's set of good manufacturing values and principles. This values-based approach provides the guiding principles for the entire organization. These are principles that can help a company make effective decisions under every circumstance imaginable. As we have read in this book, the real art to the manufacturing discipline is the ability of an organization to make effective decisions in an ever-changing environment. The industry itself continues to evolve as a result of competitive pressures. Pressures from offshore competition, for example, are currently forcing US industries to find new and better ways to compete. These competitive pressures translate into challenges throughout an organization. The successful company develops strategies to effectively handle these challenges, and Fast Cycle Production provides an organization the answers to many of these challenges.

The magic of the Fast Cycle Production approach is in its unique ability to recognize the Art, as well as the Science, of Manufacturing. The "Art and Know How" within a manufacturing organization is valued by the FCP approach. The Art that is applied to this body of science that we all know as Manufacturing comes from the vast amount of empirical knowledge that is required to effectively run a manufacturing organization. The knowledge of our factory experts has become so important to the success of a manufacturing company that it needs to be recognized, appreciated and exploited in order to give a company its strategic advantage. This extremely competitive and challenging manufacturing industry is requiring that certain disciplines and strategies be employed in order to preserve the Art as well as the Science of Manufacturing. Fast Cycle Production addresses this challenge on several fronts.

The ongoing management of empirical production data through the use of Fast Cycle's TAKT Management process is one example of how FCP differentiates itself from other popular productivity techniques. The application of Run Rules throughout the FCP factory is another example of the emphasis that FCP places on preserving the "art and know-how" that our factory experts apply to the decision making process. Changes occur every day on the manufacturing floor, and the application of Run Rules is just another example of how Fast Cycle Production helps an organization deal more effectively with change. The team approach is ever-present throughout FCP, and is employed across its various applications. The team approach helps an organization utilize the knowledge and expertise that resides within an entire organization. Developing strong leadership, promoting teamwork and forming effective partnerships within an organization are the principles upon which all ongoing improvement efforts, and Fast Cycle Production in particular, are based.

Fast Cycle Production bases its powerful manufacturing strategy on its strong set of values and principles. It is these values and principles that create the foundation for FCP. All strategies, decisions and productivity methods are rooted in these modern manufacturing values. These fundamental values involve a strong focus on delighting the customer by understanding exactly what they need and providing them precisely what they need, when they need it. FCP accomplishes this objective with a combination of customer focus, and continuous process improvement. Eliminating waste from our manufacturing processes is the primary emphasis of the FCP value-set, and the Fast Cycle Production approach requires that partnerships are forged throughout an organization in order to effectively achieve this goal.

Many organizations are discovering the benefits of a values-based management approach to its business. Companies are finding that focusing the entire team on a common set of values can enhance its organizational effectiveness. A values-based approach provides an organization with a set of guiding principles that can keep the entire company focused during good times and bad. This value-set becomes like the North Star is to a ship that has set sail on the open seas, providing an unwavering, constant set of direction which helps guide the ships journey through both calm and turbulent waters. The Values-based Manufacturing approach employed by FCP provides that constant set of direction to the entire organization. Fact Cycle Productions continues to achieve astonishing results by focusing the entire company, and particularly the leadership staff, in a team oriented, continuous improvement process, which is rooted in a strong set of good manufacturing values and principles.

While Fast Cycle Production has evolved over the years, the set of Good Manufacturing Principles has never changed. Enhancements have been made to FCP in order to respond effectively to the dynamic challenges of the manufacturing industry. An enhancement like Dynabans, for example, is an important improvement over the traditional Kanban approach that makes Kanban Management much more effective, while the principle of controlling the amount of WIP on the manufacturing floor is preserved. Future enhancements to FCP are inevitable, and while some change will be required, its fundamental values must never change. For example, Quality Functional Deployment is currently being evaluated for its applicability to FCP's Value-add Analysis process. Our vision is to utilize QFD as a more scientific approach to gather and understanding customer requirements. Utilizing the "voice of the customer" to identify the customer's requirements should only serve to strengthen the value-add analysis process. The principle of efficiently and effectively providing value to the customer is preserved, and hopefully value-add analysis will be more effective as a result of this FCP process change. The current evaluation of the QFD methodology, not an evaluation of QFD per se, as it has proven itself a very useful requirement assessment technique. The evaluation is simply an assessment of how well it integrates with the FCP project implementation process. One of the advantages of FCP is its "blue print" for implementation. Its progressive implementation process by way of a relatively straightforward project management approach allows the Fast Cycle process to be implemented very effectively in a team setting. The QFD technique must improve the results of the implementation before it is considered as an FCP offering.

Other areas under investigation include the 5 S's and the Kaizen Blitz approach to improve workflow and work area organization.

307

These areas are viewed as nice compliments to the other productivity techniques prescribed by Fast Cycle Production. In fact, the "spaghetti diagrams" used during kaizen or a Kaizen Blitz is almost exactly the same technique as the Circles of Motion methodology that Tom Clason developed almost 30 years ago. Tom developed Circles of Motion while working in a mobile combat unit during the Vietnam War, and found it to be a powerful cycle time reduction technique for his unit. Suffice to say, that the effective use of this technique is more than a business imperative to Tom, as its usefulness literally meant the difference between life and death. The principle of improving workflow, and eliminating waste from a work place albeit a workstation or department, are the fundamental FCP values that need to be observed and realized within the manufacturing process. Circles of Motions, work place organization and product flow improvements are all fundamental FCP principles; so any opportunity to be more effective in these areas are worthy of inclusion in the Fast Cycle Production approach.

Any and all other improvements or supplements to the Fast Cycle Production continuous improvement process will be put through a similar type of scrutiny. Any "new" productivity approach that does not support a fundamental principle or value, and does not achieve results in an effective manner, simply will not be considered as a Fast Cycle Production offering. Preserving the FCP value-set is critical to the success of the Manufacturing organization, and is even more critical to the Fast Cycle Production Company. The values we profess to be important for our clients are diligently practiced within the FCP organization, and will continue to be practiced throughout its existence.

About the Author

Tom Clason has spent over three decades perfecting a manufacturing philosophy and implementation plan that always produces outstanding results. His work with process excellence first started during the Viet Nam War when, as a member of a mobile tactical combat team, he first discovered that the amount of time that it took to set up or tear down a mobile base camp could be reduced dramatically by following a concept that he had developed. The concept was adopted and used by his unit with amazing results. He later named this concept Circles of Motion and has used it in virtually every one of his engagements. Circles of Motion is only one of the outstanding productivity tools developed by Mr. Clason and explained in his book.

After leaving the Air Force, Mr. Clason worked for ten years as a manufacturing manager, during which time he became a student of the accomplishments of Toyota and their use of Just In Time (JIT) concepts and tools. He became an advocate and teacher of both JIT and CFM (Continuous Flow Manufacturing). The Fast Cycle Production productivity improvement system of today is foundationally based on JIT and CFM principles.

In 1989, while working for IBM, Mr. Clason joined a pilot external consulting organization that would eventually become the IBM Consulting Group. After five successful years with the IBM Consulting Group as a productivity/simulation modeling consultant, during which time he was 90% billable; Mr. Clason left IBM to become the Owner/CEO of Fast Cycle Production Consulting. His success with helping companies to improve their processes not only continued, it reached new highs.

During his work as a consultant, Mr. Clason's client list both with IBM and FCP is a "Who's Who" of Manufacturing, as well as some very influential and successful service organizations such as Bank of America and Colorado Springs Utilities. His client list includes manufacturing companies such as, Ethicon, Sikorsky, Baxter Health Care (Now known as Allegiance), Keytronics, Engelhard and Rock Tenn. His system has proven to be highly effective in virtually every industry. The fact that his Values Based productivity improvement system has never failed to produce outstanding results is still his proudest accomplishment.

www.ingramcontent.com/pod-product-compliance
Lightning Source LLC
Chambersburg PA
CBHW031822170526
45157CB00001B/148